Dedication

In honor of the United States Air Force's 50th anniversary, this book is dedicated to the men and women who serve America either in uniform or as civilian contributors to our national security and global economic strength, through cooperative technological exchange with the research, development, test, acquisition, and support communities.

TECHNOLOGY EXCHANGE IN THE INFORMATION AGE

A Guide to Successful Cooperative R&D Partnerships

Second Edition revised by

JOHN LESKO
PHILLIP NICOLAI
and
MICHAEL STEVE

BATTELLE PRESS
Columbus • Richland

Disclaimer

This book was prepared as an account of work sponsored by an agency of the United States Government. Neither the United States Government nor any agency thereof, nor Battelle Memorial Institute, the Economic Strategy Institute, nor any of their employees, makes any warranty, expressed or implied, or assumes any legal liability or responsibility for the accuracy, completeness, or usefulness of any information, apparatus, product, or process disclosed, or represents that its use would not infringe privately owned rights. Reference herein to any specific commercial product, process, or service by trade name, trademark, manufacturer, or otherwise does not necessarily constitute or imply its endorsement, recommendation, or favoring by the United States Government or any agency thereof, or Battelle Memorial Institute or the Economic Strategy Institute. The views and opinions of the editors and authors expressed herein do not necessarily state or reflect those of the United States Government or any agency thereof.

Supportability Investment Decision Analysis Center
operated by
Battelle Memorial Institute
for the
United States Air Force Materiel Command
under Contract F33657-92-D-2055/0036

Library of Congress Cataloging-in-Publication Data

Technology exchange in the information age : a guide to successful cooperative research and development partnerships – 2nd ed. / rev. by John Lesko, Phillip Nicolai, Michael Steve, Jr.
p. cm.
First ed. published under title: Technology exchange. c 1995.
Includes bibliographical references and index.
ISBN 1-57477-037-3 (pbk : alk. paper)
1. Technology transfer—United States. 2. Research and development contracts—United States. 3. Information technology. I. Lesko, John (John N.), 1957- . II. Nicolai, Phillip (Phillip A.), 1943- . III. Steve, Michael (Michael L.), 1948- .
T174.3.T3754 1997
338.97307–dc21 97-30038
CIP

Printed in the United States of America

Battelle Press
505 King Avenue
Columbus, Ohio 43201-2693
614–424–6393; 1–800–451–3543
Fax: 614–424–3819
Homepage: http://www.battelle.org/bookstore
e-mail: press@battelle.org

Contents

Foreword . . . to First Edition

The leaders of the 21st century will be defined not only by military might, but more importantly by their technological, economic, and knowledge prowess guided by a coherent economic strategy. As resources continue to diminish, both industry and government are seeking ways to leverage their investments in technology. Cooperative research and development between the public and private sector is one leveraging tool that holds considerable promise.

The editors and contributing authors of this book have conducted extensive, original research into cooperative technology programs. Their findings demonstrate that, properly executed, cooperative technology programs benefit both industry and government, through enhancing the research, development, and application of critical technologies.

Industry benefits through access to new technologies and reduced R&D costs. For the federal government, these programs hold the promise of support for existing systems, while ensuring the fielding and supportability of future systems. Central to future success in this critical undertaking is joint strategic planning between industry and government for the development and deployment of critical technologies.

The Economic Strategy Institute is pleased to join with Battelle and the Supportability Investment

Decision Analysis Center (SIDAC) in this endeavor aimed at advancing the understanding of the unique challenges and rewards presented by industry/government cooperation.

—Clyde V. Prestowitz, Jr.
President, Economic Strategy Institute

Foreword . . . to Second Edition

Much of the investment in R&D work in the period from 1960 to 1990 was driven by a quest for fundamental knowledge and a rapidly expanding and changing economy. Scientific and technological developments were pursued, with the result that a vast platform of knowledge was made available just in case it was needed. In recent years, the changing attitudes and financial pressures (at both the government and industrial levels) created a fundamental change in posture, one which evolved into more of a "just in time" approach: take the technology that is available from whatever source and modify/implement as necessary in order to accomplish the objectives (usually short term) of the near-immediate future. Funding for "just-in-case" knowledge was replaced with funding for "just-in-time" application. This approach has permeated almost every sector of the economy.

As we move into the new century, the emerging set of realities is different from what the past researchers have experienced. The old standards—increased R&D budgets, basic research concentrations in industry, long-term outlooks toward technological progress, and the continued real growth of technology-based opportunities throughout the U.S. economy—will be replaced. New relationships within the overall R&D enterprise—both domestic and foreign—will have to be defined and fostered. New and expanded ways of doing business will have to be proposed, designed, and endured. And, as a corollary, old barriers will have to be addressed and overcome. In this context, two-way technology exchange and transfer has the potential to become the centerpiece for new actions.

The entire concept of technology transfer has now emerged with a different understanding of the needs and opportunities of both sides of the technological fence. Both industry and the government laboratories need each other for effective technology flows that support each others' missions. The federal laboratories need to develop a broader clientele—while maintaining the capabilities to respond to the first-line needs of their mission agencies. Industry needs sources of technology which augment those that are available from other resources. And both sides need the economic and political support required to provide continuity, reduce risks, and capitalize upon the applications of technology. There is little doubt that there are both needs and opportunities on each side of the technology transfer/technology receipt interface: the major task lies in the identification of the right type of information at the right time and involving the right people. *Technology Exchange in the Information Age* helps show the way.

—DR. JULES DUGA

Preface . . . to First Edition

Among our most pressing imperatives is the need to speed the application of research and development between the government and the private sector in an ongoing and strengthened exchange of technology. Government investments in research and development need to be leveraged with private sector investments to make maximum contributions to the nation's economic and military strength. From a military perspective, success will depend upon a science and technology investment strategy that develops, deploys, and supports the best possible systems required to fight and win every future conflict. From an industry view, this strength and willingness to participate comes from a national policy that encourages and facilitates the necessary critical involvement of American industry in the global technology marketplace.

Technology Exchange: A Guide to Successful Cooperative Research and Development Partnerships is about making technology transfer a two-way exchange between the public and private sectors. The focus of this guide is on two essential components of national security, namely U.S. military preparedness and American industrial competitiveness. It has particular relevance for technologists, business managers, leaders of federal and state governments, regional organizations, academic institutions, and commercial and defense industries—both large and small—who view the potential of cooperative research and development (R&D) partnerships as a possible component of their investment strategy. The net effect of this two-way exchange will be a a "win–win" strengthening of our global competitive position and military/industrial base.

Technology Exchange is the product of a team effort, sponsored by the United States Air Force Materiel Command (AFMC), to enhance the effectiveness of the Air Force technology transfer program. The results, although certainly reflecting the needs of the science and technology community within the Air Force, should benefit anyone involved in or planning to be involved in technology transfer partnerships. The overall project called for a critical examination of the existing technology transfer process, in particular cooperative research and development activities and the pursuit of dual use technologies. The research plan included a thorough review of the literature, a joint survey of public and private sector participants who have entered into cooperative ventures, and benchmarking visits to both governmental and industrial facilities to learn directly from those who have demonstrated success. The effort culminated in a series of roundtable discussions for leaders to address a broad range of issues. The guidebook itself was carefully developed from the results of each of these project elements. It is intended to assist both government and industry participants.

This book is offered to encourage both government and industry participants in the various technology transfer and cooperative exchange activities. We believe that it will prepare both parties for the challenges presented by technology transfer, and help ensure that participation will be mutually beneficial to all. In this book, we present a candid assessment of the technology transfer community today—good and bad—so that all parties may enter into partnerships with realistic expectations and their ears and eyes wide open.

The main body of text in *Technology Exchange* begins with an overview in Chapter One (*Government/Industry Cooperative R&D: Is It Right For You?*) that establishes the importance of technology transfer activity and joint planning in leveraging our federal technology assets for national security purposes as well as in optimizing the potential payoff for your organization. We make the argument that technology transfer must be fully integrated into your investment strategy. In Chapter Two (*The Present Environment for Technology Partnerships*) we outline some widely held perceptions on the uses of cooperative R&D and the challenges of developing an R&D agreement between partners from the private and public sectors. Then in Chapter Three (*Cross-Cultural Considerations/Imperatives*), we examine the major barriers or impediments to successful cooperative R&D. These impediments are described in cultural terms, because most serious problems stem from differences between the public and private sectors through their traditional business policies, practices, and imperatives. The basis of these findings comes from an extensive survey of CRDA participants and from our benchmarking effort.

Chapter Four (*Building Bridges*) begins the constructive process towards making cooperative R&D work better. It discusses ways to improve communications, enhance organizational focus and accountability, acquire outside help, and educate the participants at every level, all of which are actions that help bridge the cultural gap described earlier. In Chapter Five (*The CRDA Process*) the research team discusses the planning that goes into building a successful partnership agreement, then concentrates, in some detail, upon the CRDA instrument itself and the negotiable items of a cooperative R&D agreement. The chapter concludes by introducing relevant vignettes of "best practices" and "lessons learned" to illustrate solutions that have worked for many organizations. Chapter Six (*How to Achieve Success*) describes the role and importance of key performance elements, including incentives, metrics, leadership and strategic investment planning. It also acknowledges that successful partnerships start with and depend upon world class technology and a "make it work" attitude.

The main body of the text ends with the *Summary*, which highlights the best practices and lessons learned, and enumerates the characteristics that go into making technology transfer a "win–win" experience.

Important supplemental materials are provided in the back matter to this book. Appendix A gives the Executive Summary of *Play to Win*, the concept paper written by the Economic Strategy Institute that started the Air Force study of ways to improve their technology transfer program. Appendix B provides information on both the organizations that made up the research team and brief biographical sketches of team members participating on the actual technology transfer study. Appendix C gives a summary of Technology Transfer Legislation and Executive Orders, listing the year of enactment, name, and major elements of the legislation as passed. Appendix D provides a glossary of terms used in descriptions of the technology transfer process, and Appendix E is an overview of technology transfer mechanisms, giving definitions and describing each mechanism's chief characteristics and strengths. Appendix F lists the more popular and useful bridging organizations. It provides an overview of each organization's program and offers key points of contact. Appendix G gives the text of a model Air Force CRDA and includes a checklist for ensuring that all relevant issues are addressed in negotiating the agreement. The *Bibliography* provides a list of the published resources drawn upon in the analyses behind *Technology Exchange.*

Technology Exchange provides a provocative insight into the world of technology transfer and to the benefits and challenges associated with cooperative R&D. By using the complete set of government furnished documents, laws, handbooks, etc., and by complementing these with "best practices" and "lessons learned" described in this guidebook, we think the technology transfer process in general, and the success rate of future CRDAs specifically, will be improved.

—ANDREW DOUGHERTY
Senior Fellow, Economic Strategy Institute

—PHILLIP NICOLAI
Senior Research Scientist, Battelle/SIDAC

Preface . . . to Second Edition

Technology Exchange in the Information Age is an updated and revised edition of *Technology Exchange: A Guide to Successful Cooperative Research and Development Partnerships*. The Internet, the World Wide Web, on-line information sources, information analysis centers, subscription-based information providers, and other "new media" references have significantly changed the environment for cooperative R&D. Our original study was commissioned by the Air Force Materiel Command Science and Technology Directorate in 1993. Our first manuscript was published in 1995. A lot has changed within the R&D community since then.

In the last two years, the Internet has seen phenomenal growth. The precise number of new users is hard to determine—statistics range from 12 to 26 million users in the US—depending on which source you cite or believe.* If you look at these numbers more closely,

*The readers should check for themselves the vast statistics cited here by going to the Yahoo search engine (URL: http://www.yahoo.com) and "surfing" the category Computers, Internet, Statistics. For example, Matthew Gray at the Massachusetts Institute of Technology reports, "that there are 9.5 million host computers on the Internet as of 1996." O'Reilly & Associates, an oft cited Internet research firm, projected in October 1996, "that there will be 15.7 million users on-line by the year's end." Nielsen Media Research claims, "24 million Internet users, 18 million WWW users, [and that] users average 5 hours and 28 minutes per week on the Internet."

you find that the majority of Internet users are well educated, technically savvy, and using on-line resources in ever increasing ways to do their day to day jobs. Twenty-five percent of Internet users come from engineering (15%) and R&D (10%) disciplines. The information age is in full swing, particularly in the laboratories and engineering centers of the nation. Concurrent with this expansion in Internet use has been the contraction of the Federal investments in defense-related R&D. Practitioners in technology transfer have had to rely more and more on commercial sources of R&D know how. Growing experience within a broadened defense technology and product oriented community, and new policy/legislation are paving the way for a more mature technology exchange process and a greater "win-win" incentive to participate.

This new publication, *Technology Exchange in the Information Age*, preserves (and to a great extent revalidates) the results of our earlier research. Most relavent is the central finding emphasizing the mission-enabling, two-way nature of technology transfer and cooperative R&D. Follow-up surveys and interviews were prevalent techniques used throughout our continuing study effort with participating practitioners and managers from the scientific and technical information and technology transfer communities. However, original research for this new edition was primarily concentrated in three principal areas, namely: (1) the US military and federal budget climate and the ongoing macro economic shift from public to private sector R&D funding; (2) revolutionary advancements with the Internet, providing greatly enhanced information brokering capabilities realized through self-help browsing and more accessible bridging activities; and (3) new policy/legislation, in particular the National Technology Transfer and Advancement Act of 1995.

New and updated material reflecting each of these major changes has been appropriately integrated throughout the manuscript into the basic structure and text of the earlier edition, with one new appendix (Appendix H) on technology transfer related Internet sites.

—PHILLIP NICOLAI
Senior Research Scientist, Battelle/SIDAC

Acknowledgments

. . . to First Edition

Technology Exchange is the principal product of a year-long study entitled: "Improving the Air Force Technology Transfer Program—Best Current Practices and Alternative Approaches to Outstanding Issues." The study was conducted for the Air Force by technical engineers, research scientists, economic analysts, systems and program managers, and supportability experts from Battelle and the Economic Strategy Institute (ESI). Their dedication, professionalism, and commitment to quality research made it possible to extend the initial Air Force study findings to a broader audience interested in technology transfer.

We wish to acknowledge the ESI team members responsible for research and analytical assistance, who authored major segments of this book. They include Amy Belt, Morgan Fargarson, James Hall, Michael Kull, Tedd Ladd, Brian Wagner, and David Weinstein.

We wish to express our deep appreciation to three

special individuals: Andrew Dougherty, Phillip Nicolai, and Michael Steve. These three professionals championed our work. Without their support, this guide would have never materialized.

Andrew Dougherty, as Economic Strategy Institute project director, provided the energy, learned counsel, visibility, and excellent control over the Economic Strategy Insitute team activity, and offered his understanding of the business and policy sides of this process. His efforts included recruiting world-class advisors and participants for every phase of the project.

Likewise, we would like to thank Phillip Nicolai of SIDAC and Battelle Technical Services Organization located in Dayton, Ohio. He provided overall task leadership, understanding of the key roles that mission and systems support play for the government worker in this technology partnering process, and help with the written word as a major contributing author.

Michael Steve's assistance was invaluable in bringing together the constructive criticisms and remarks of more than two dozen reviewers of a draft version of this guide. The discipline he brought to the team was invaluable.

We would like to acknowledge the outstanding leadership provided by Major General Richard Paul, AFMC Director, Science and Technology, for the vision and foresight to seek solutions and for assuring much appreciated government participation in those activities critical to every phase of this effort.

We would like to thank our Air Force sponsor, Eric Werkowitz of Headquarters AFMC, Science and Technology Directorate, for his patience, valued inputs, support, and encouragement. Tim Sharp, AFMC Technology Transfer Office, offered valued commentary throughout this project, and the Air Force's Directorate for Science and Technology supplied much of the material that appears in Appendices C and G.

We are particularly indebted to several hundred willing and dedicated participants in the survey, benchmarking, and round table segments of the project; without their sacrifice of valuable time, we could not have achieved our goals with this effort. Valued input during the survey phase of the Technology Exchange study was received from Dr. William Beusse of the U.S. General Accounting Office's National Security and

International Affairs Division. Mark Blazey served as facilitator of all roundtable discussions with representatives of government and industry. Preparation for these discussions and advice for our book came from many professionals. We would like to make note of the assistance we received from Dr. Roy Amara (Strategic Decisions Group), Dr. Beverly Berger and Dr. Loren Schmid (both with the Federal Laboratory Consortium), and Dr. Barry Bozeman (Georgia Institute of Technology). Robert Widder (Battelle) offered key insights into the survey and benchmarking phases of the study. Of the more than two dozen reviewers who provided comments, we single out for acknowledgment Tina Macaluso McKinley (Chair, Federal Laboratory Consortium), Jacob "Jesse" N. Erlich (Chief Patent Advisor for the Electronic Systems Center, U.S. Air Force), and George Krikorian (Professor of Program Management and Industry Chair at the Executive Institute of the Defense Systems Management College); their cautions, corrections, and reactions were used to reinforce the text and supplemental materials.

Finally, we must point out that no federal employee has been compensated in any way for contributions to this work. The responsibility for any short-comings, errors of omission, or other inaccuracies rests solely with the editors. We have benefited greatly from working with literally hundreds of participants in this project and are indebted to them all.

THE EDITORS
John Lesko and Michael Irish

Acknowledgments . . . to Second Edition

In the last two years since the publication of *Technology Exchange: A Guide to Successful Cooperative Research and Development Partnerships*, various practitioners of technology transfer have formed a kind of relay team, collecting and handing off materials that have contributed to this new text. These include Battelle, the Supportability Investment Decision Analysis Center (SIDAC), select members of the US Air Force, and a network of technology transfer consultants. This expanded network includes all those who participated in the Air Force's Scientific and Technical Information (STINFO) Conference, several technology transfer workshops by the Affiliate Society Council of Dayton, and the SIDAC-sponsored conference held in Dayton, Ohio, on the topic of "STINFO and its role in the technology transfer and commercialization process."

The nucleus of our research team, listed alphabetically and in no direct relationship to their contribution, includes: James Kelly, John Lesko, Phil Nicolai, and Michael Steve. This team worked as interviewers, researchers, facilitators, and participants at the above named conferences validating and adding to the "lessons learned" and "best practices" presented in the first edition.

We would like to make some special acknowledgments. Mike Silverman from the Air Force's Wright Laboratory served as the sponsor for the continuation study. Pat McWilliams, Bill Whalen, Chuck Chatlynne, Carlynn Thompson, Michael Schrage, George Krikorian, Roland Gonano, Jim Singer, Tim Sharp, John Boeck, Dr. Jules Duga, Major Richard Franza, Dr. Charles Herzfeld, LtCol Larry Kosiba, and Jeff Krattenmaker served in various advisory capacities or as panelists at the Dayton technology transfer seminar and workshop, or both.

Finally, as in the first edition, we must point out that no federal employee has been compensated in any way for contributions to this work. The responsibility for any short-comings, errors of omission, or other inaccuracies rest solely with the editors. We have benefitted greatly from working with all who have participated in this revision and continue to be indebted to them all.

THE EDITORS
John Lesko, Phillip Nicolai, and Michael Steve

TECHNOLOGY EXCHANGE IN THE INFORMATION AGE

A Guide to Successful Cooperative R&D Partnerships

chapter 1

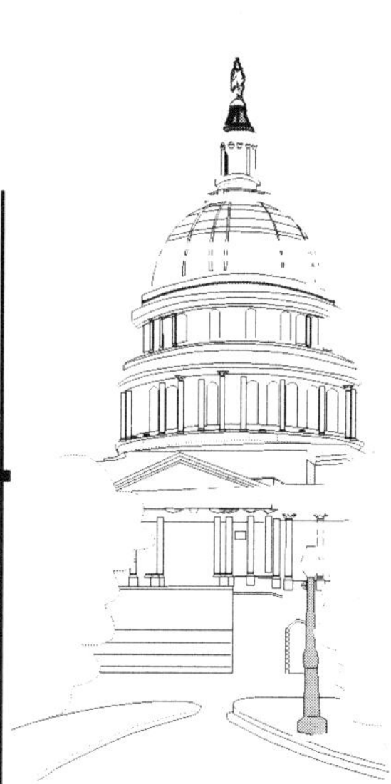

Government/Industry Cooperative R&D: Is It Right For You?

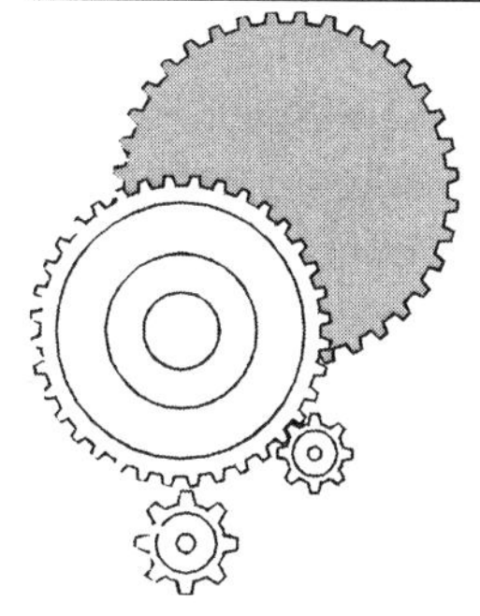

It is the continuing responsibility of the Federal Government to ensure the full use of the results of the Nation's Federal investment in research and development. To this end the Federal Government shall strive where appropriate to transfer federally owned or originated (non-classified) technology to State and local governments and to the private sector.

— Excerpt from the Stevenson-Wydler Technology Innovation Act

The commercialization of technology and industrial innovation in the United States will be enhanced if companies, in return for reasonable compensation to the Federal Government, can more easily obtain exclusive licenses to inventions which develop as a result of cooperative research with scientists employed by Federal laboratories.

— Excerpt from the National Technology Transfer and Advancement Act of 1995

AN OVERVIEW

Fundamental changes are taking place in economic and political systems around the world. Old adversaries are showing an inclination to embrace concepts of democracy and a free market system. New adversaries are redefining the areas where their and our national interests collide, and the actions that constitute a challenge to those interests. As a result, policymakers and business leaders alike are attempting to reevaluate what the strategic interests of the United States are, and determine what role the United States should have in the world. A comprehensive coherent national security strategy must address tomorrow's uncertainties, military and, increasingly, economic.

In the coming years, the United States will face growing competition as it seeks to protect, strengthen, and facilitate its interests around the world. We already recognize the economic strength of Japan and the European Community, which have adopted national strategies which acknowledge the need to, and embody the ability to, coordinate and strategically deploy public and private sector resources in a focused manner during peacetime. The results include impressive industrial, technological, and economic successes such as the Japanese automobile industry and Airbus produced by the European aircraft consortium.

Our chief competitors have national strategies to coordinate and deploy public and private sector resources.

Competition is also coming from nations less well off than those of the European Community or Japan. Several newly industrialized countries are demonstrating a capacity to pool resources in order to bolster their economic might and achieve regional advantages. Developing nations, such as China and Malaysia, have adopted the Japanese economic model, embracing high technology, high quality, production-oriented values, and the corresponding seamless cooperation of the public and private sectors.

In the face of well coordinated, broad-based competitive challenges, we as a nation must recognize the close linkage between economic strength and national security. Global realities dictate that we develop a strategy that maintains our nation's technological superiority—the linchpin of our economic and military security—while aggressively pursuing and stimulating new commercial industries. In many cases, the key to success lies in nurturing and protecting critical technologies and industries.

We must recognize the close linkage between economic strength and national security.

The laboratories under the Department of Energy, Department of Defense, Department of Commerce, Department of Agriculture, National Institutes of

Health, NASA, etc., have been designed to meet national challenges through basic and applied research. Billions of dollars have been invested through the decades to maintain U.S. superiority in research, development, and engineering. These technological knowledge assets, made available through technology transfer programs, can play an integral role in improving our national competitiveness. Equally important, exchange of technology based on a dynamic partnership can play a vital role in maintaining needed technological superiority and capabilities for the federal missions they support.

Back when plenty of federal R&D and procurement dollars were available, any spin-off application of DoD technology was considered a "bonus," i.e., getting extra value from the R&D dollar by strengthening the civilian economy. Such technology transfer was a one-way flow from the separate defense technology and production base to the separate commercial technology and production base. This arrangement, which we call the old paradigm, was appropriate when the U.S. faced a monolithic threat that emphasized military competition.

Now the concept is a two-way exchange of technology capabilities, opening the way for spin-on as well as spin-off applications. The goal is a seamless integration of military and commercial capabilities so we as a nation can maintain military capability and develop emerging technologies. In areas where military R&D has taken the lead, this technology must be transferred for two reasons: to afford the material that supports day-to-day operations, and to strengthen the domestic economic base. There are only two areas where a separate military technology should be maintained: (1) technology that is unique to military application, and (2) technology deemed critically sensitive. Military acquisition and logistics programs outside these critical areas can benefit from the application of successful business practices. The opportunities for technology transfer to the military will grow as defense program managers are allowed to establish and pursue performance specifications that are more current than MILSPEC, and obtain products, systems, and other deliverables from an integrated production line in the private sector. This new paradigm of technology transfer is appropriate now because we have to deal with predominantly economic

The new paradigm is a seamless integration of technology capabilities across the public and private sectors. Recent reform initiatives take a step in this direction by allowing defense program managers to apply proven business practices to military acquisition and logistics programs.

competition, because the military threat is diminished, less immediate, and more diverse.

Technology transfer serves as an interim step to what may eventually resemble an integrated military and commercial industrial base. The foundation has been laid through policies and public law that have been adopted to stimulate technology transfer and cooperative R&D between the public and private sectors. Defense conversion, dual-use technology, technology transfer, and cooperative R&D are all policies and programs that ultimately are intended to bring the public sector closer to the private commercial sector.

From a business perspective, the environment for government/industry cooperative R&D has never been better. Whether you represent the interests of a private enterprise or a federal institution engaged in R&D activities, evolving market realities portend the need for a new approach to, or at least a fresh look at ways to achieve corporate goals for improved technology in a tighter, more competitive national and international market environment. Faced with dramatic changes in mission and drastic cuts in resources on the one hand, and ever increasing demands for leading technology in the programs that remain, the federal program director or technologist must look beyond the traditional defense contractor community and consider solutions that may be offered by the private sector. Similarly, pressures to reduce corporate R&D budgets make the commercial entrepreneur more inclined to investigate public sector possibilities for cooperatively leveraging at least a portion of higher risk or costly early development work.

If you are, in fact, in a position to effect or implement change in your organization's technology investment strategy that includes greater participation in cooperative R&D activities, then a closer examination of the concept of integrated strategic planning is warranted. This topic is important not only from a national perspective in leveraging critical industries and technologies, but also from an organizational perspective in incorporating the potential for *Technology Exchange* into a vision for corporate success and military mission success.

THE NEED FOR AN INTEGRATED STRATEGY

Two ingredients are essential to make a two-way technology exchange work. The first, joint strategic planning, is consummated in our national policy and legislation, and provides a framework for exchange. It nurtures an improved environment for cooperative R&D, by helping to replace an "arms-length" relationship that has persisted due to the traditionally independent roles of government, industry, and academia in conducting and reporting S&T activities. The second ingredient, integrated technology investment planning, provides the real impetus to implement change. A technology investment strategy that recognizes the potential for technology transfer and cooperative R&D, and fully integrates it into the organization's business process, can leverage scarce resources, broaden knowledge, and increase the chances for success.

Joint strategic planning improves the prospect of technology transfer between the public and private sectors.

Supporting the weapons systems that will ensure military success is directly dependent on technology superiority that has emerged from the nation's federal, commercial, and university laboratories. Innovations and technologies that characterize the "information age" are becoming dual-use, with potentially broad application to commercial and military requirements. Increasingly, products and technology are appearing first in the commercial market, to be adapted later to military needs. A strong domestic technology base is guaranteed through government and industry cooperation, which formalizes a commitment to manage limited resources better and drives new technology innovation in both sectors.

Government-generated technologies can have commercial value if developed further by the private sector.

Many technologies and techniques generated in the federal laboratory system can have commercial value if developed further by the private sector. This perception drives the policy and efforts to improve the technology transfer process. Ultimately, technology transfer and cooperative R&D can lead to joint strategic planning between the public and private sector, enabling more cost-effective future procurements.

The term "technology transfer" is a misnomer, since there are few instances of technological innovations that are ready to be moved from the shelves of the federal laboratories to the shelves of the commercial mar-

Innovations do not simply move from the shelves of federal labs to the shelves of the commercial market. Most require extensive and expensive modifications, and even then there is no guarantee of success in the marketplace.

ket. In fact, of the technologies developed by the federal laboratories that are not military-specific, most are general in application feasibility, and require extensive and expensive modifications before they can be packaged for market. The combination of a sound technology investment strategy and joint research and development planning provides the occasion for the federal laboratories and private sector laboratories to enhance American competitiveness by developing dual-use technologies. Cooperation could lead to improved development processes, integrating both design and production so that the work done in an Air Force laboratory, for example, might be transferred more easily to industry.

Technology transfer and cooperative R&D are no longer in their infancy. Policy guidelines and expectations are more realistically agreed upon by organizations that have been practicing partnerships from the start of these programs. However, the overall maturity of most tech transfer programs is still at the "toddler stage." For organizations getting started and for many having just learned to walk, initiatives are still characterized by growing pains and process impediments—cultural, legal, bureaucratic, and financial—that can be difficult to deal with. However, more industry and government organizations are coming to realize that technology transfer and cooperative R&D can support their goals and missions. This, in turn, is increasing the level of energy being applied to overcome those impediments.

An improved technology transfer process would greatly reduce the delay industry experiences in trying to commercialize a technology developed by the federal laboratories. Eliminating delays is important, because great economic benefits accrue to the first to commercialize a new product. America, therefore, must implement innovative policies that will enhance her ability to compete in uncertain geopolitical scenarios and with rapidly evolving economic systems. The authors of this study believe that the technology investment strategy of the federal laboratories and partnering organizations must include policies that enhance the technology transfer process to ensure the continued economic strength of our nation in an increasingly competitive global market.

TECHNOLOGY TRANSFER AS NATIONAL POLICY

Over the past five decades, the defense industrial base of the United States has spawned the aerospace industry, among other successes, and has served as the nation's technology seed bed. As the defense industrial base contracts, and public and private resources become more scarce, a merging of the commercial and defense industrial base becomes imperative if we are to sustain national technological core competencies. Technology transfer and cooperative R&D provide an attractive, cost-effective, well-leveraged alternative to independently funded research and technology development. The sharing of knowledge, resources, and innovation in transfer of technology and cooperative R&D among the public sector, defense contractors, academia, and other organizations has become a necessity.

The annual federal investment in R&D is approximately $70 billion, of which roughly 60 percent is defense specific. For the federal government, cooperative R&D programs can reduce duplication of research efforts. For industry, cooperative R&D programs can open the door to government resources, technology, equipment, and expertise. Because the federal R&D base is broad, technology transfer programs offer potential benefits to large, medium, and small firms, both inside and outside the traditional defense establishment.

The knowledge base created by various federal R&D organizations could serve as a foundation for commercially promising efforts in the private sector. For many years the federal laboratories have created research assets that embody some of the most impressive scientific talent in the world. While these research teams have historically concentrated on defense research, today, in cooperation with industry, they are focusing more of their resources on enhancing American competitiveness.

The knowledge base of federal laboratories can be a foundation for some commercially promising efforts.

Over the years, the federal government has attempted to promote the transfer of technology developed in federal research laboratories to state and local jurisdictions, as well as to the private sector. The primary law affording access to the federal laboratory system is the Stevenson-Wydler Technology Innovation Act of 1980, as amended by the Federal Technology Transfer Act of 1986, the Omnibus Trade and Competitiveness Act, the 1990 Department of Defense Authorization Act, and the National Defense Authorization Act for FY 1991. The Stevenson-Wydler Act made technology transfer part of the mission requirements of the scientific and engineering personnel of the federal departments and agencies (with the exception of the National Aeronautics and Space Administration, which already had technology transfer as a mission from its founding). Additional incentives for the transfer of technology were contained in several amendments to the Stevenson-Wydler Act. The Federal Technology Transfer Act allows government-owned, government-operated laboratories (GOGOs) to enter into cooperative research and development agreements, commonly known as CRDAs, with universities and the private sector. The authority to enter into these agreements was extended to government-owned, contractor-operated laboratories (GOCOs) in the FY 1990 Defense Authorization Act.

The mandate is to make technology available; not to assist in commercialization Commercialization is left to the private sector.

Support for technology transfer received an initial boost from President Clinton in 1993. His policy statement, "Technology for America's Economic Growth, A New Direction to Build Economic Strength," recommended that 10 to 20 percent of the budgets of certain Department of Energy, Department of Defense, and National Aeronautics and Space Administration laboratories be used to increase R&D partnerships with industry. The President also suggested that obstacles to CRDAs be eliminated and that other means to facilitate cooperation between government and industry be implemented.

The Federal Acquisition Streamlining Act of 1994 (FASA-94) was signed into law by President Clinton on October 13, 1994. Regulatory streamlining and specific acquisition reforms have resulted from the Section 800 panel and are being implemented. In addition to the FASA-94, the National Performance Review and other initiatives taken to "re-invent government" are beginning to take effect across all government agencies involved in research and development.

A second boost was provided by the National Technology Transfer and Advancement Act of 1995, which the House of Representatives passed in December 1995, and which the Senate passed with amendment in February 1996. Key features of the Act, enacted as Public Law 104-113, include guidance in defining the intellectual property rights of private sector partners for technologies created from joint research and development activities conducted in partnership with Federal laboratories. Industry partners are guaranteed, at a minimum, an exclusive license in a prenegotiated field of use for the new technology. Incentives are provided for Federal inventors to develop new inventions in their fields of research, and the Federal labs are granted greater flexibility to use royalties resulting from commercialization of Federal inventions to support the work of their laboratories and reward participants in CRDA activities for their work on successful projects.

While the enacted legislation is a mandate to make technology available, for example through licensing and cooperative R&D agreements, commercialization is left to the private sector. Contrary to a frequent misinterpretation made by the private sector, there is no mandate to assist in commercialization. With the exception of a few initiatives such as Office of Defense Programs within the Department of Energy, technology transfer and cooperative R&D activities have found uneven and limited funding. This situation is changing as the Administration and Congress pay increasing attention to the benefits of moving selected technological achievements of the federal laboratories to the industrial base for commercialization. Organizations like the Business Executives for National Security (BENS) are recommending private financing approaches as the ideal mechanisms for technology transfer and defense conversion endeavors. A privately run fund for defense conversion could organize support for individual venture capitalists, and offer low-interest, long-term loans or loan guarantees to private firms seeking to convert all or part of their defense activities to non-defense application. Increased emphasis on cooperative R&D and dual-use technology transfer offers a strategy that enables defense contractors to stop well short of complete withdrawal from the military-industrial infrastructure.

Next we briefly examine the ways in which the Federal Government is changing its organizational focus and adopting technology transfer and cooperative R&D program strategies.

THE MACROECONOMICS AND MICROECONOMICS OF FEDERAL TECHNOLOGY TRANSFER

Investment in research and development in the U.S. is expected to reach about $192 billion in 1997. This represents a real increase following years of stalled investment from the early 1990s, according to the annual Battelle/*R&D Magazine* forecast. The current trend is being more positively influenced by industrial actions, while government support for R&D continues at a level that is below previous investments. Dr. Jules Duga, principal author of the forecast noted that "the increases in R&D commitment are largely the result of industry's growing realization that the structural and operational changes of the recent past are not the only road to profitability. Continued investment in research will be required for long-term survival." Figure 1-1 gives the macroeconomic view of R&D in the U.S. today, with breakouts on the sources of R&D funding and the performers of R&D predicted for 1997.

The Federal Government will spend approximately $62.2 billion conducting R&D in 1997, about one-half percent more than was spent in 1966. The major portion of this spending supports needs in the area of defense, and some of this is for unique military applications. However, some government R&D is leading to products and processes finding direct application in the private sector. Thus, it should come as no surprise that companies would be interested in developing a competitive edge by entering into a cooperative R&D agreement with government entities. Figure 1-2 depicts the distribution of federal R&D dollars from 1988 to 1998 (projected) among major participants. The trend indicates that, on a percentage basis, the vast majority of federal government R&D funds remain in the Department of Defense.

To get a picture of the microeconomics of federal technology transfer, we start with the President's fiscal year 1998 budget, which contained a total obligational authority of $266 billion for the Department of Defense. Of this total, slightly more that a quarter

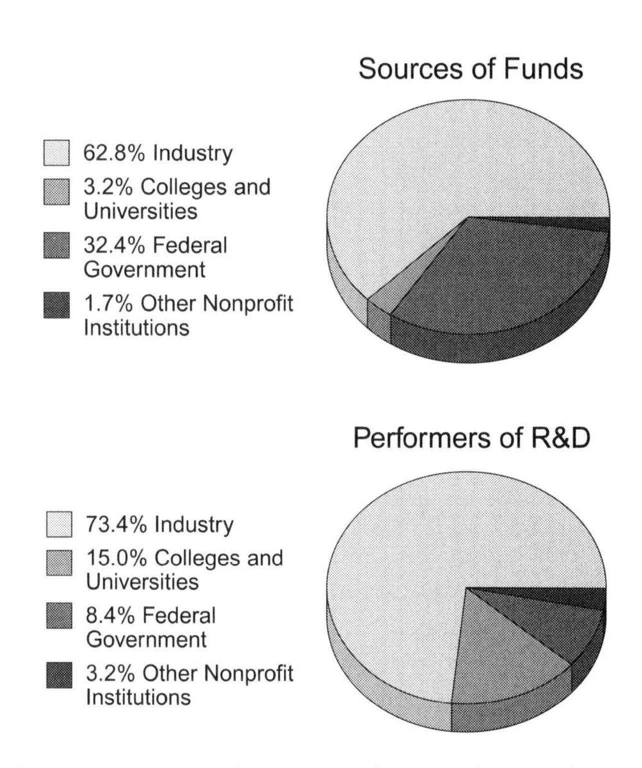

FIGURE 1-1. Projected R&D expenditures in the U.S. for calendar year 1997.

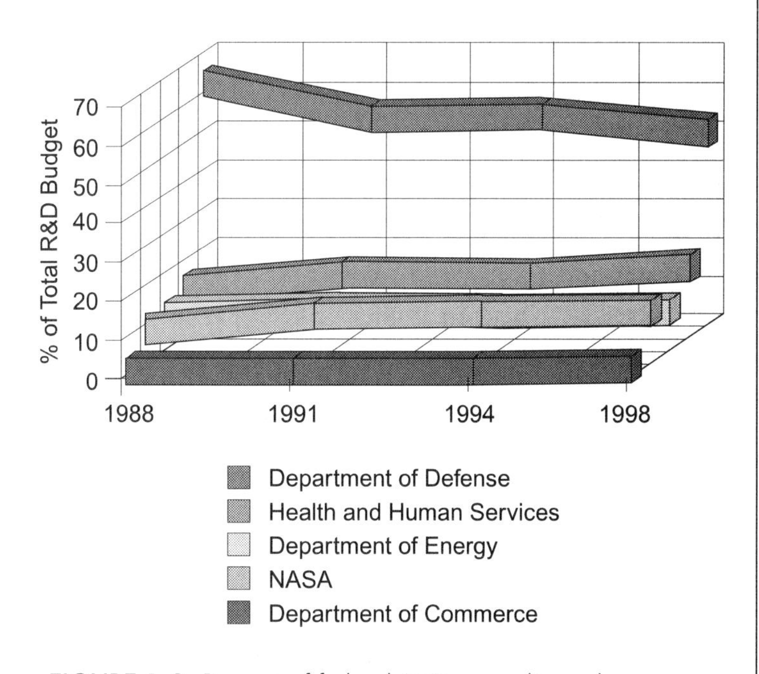

FIGURE 1-2. Percent of federal R&D expenditures by agency. (OMB data for R&D to year 1998.)

($89.9 billion) is allocated to the Air Force to cover the entire scope of Air Force business. This includes personnel costs, aircraft procurement and maintenance, base operations, training, environmental clean-up costs, research programs, etc.

The Stevenson-Wydler Act and related policies have added technology transfer to the duties of more than 4,000 science and technology staff in the Air Force Materiel Command. Other federal departments and agencies are similarly affected.

By the time we get to the accounts established for technology transfer under the provisions of the Stevenson Wydler Act and related policy directives, we are looking at a single-digit percentage of the budget. This means that there are highly leveraged dollars in technology transfer program accounts under a given command, like the Air Force Materiel Command.

The end of the Cold War has brought a new focus on improving the U.S. domestic economy. Rapid commercialization and deployment of new technologies have become an increasingly important part of the Defense Department's R&D and acquisition strategy. Both in theory and in practice, the use of commercial items provides equivalent or near-equivalent combat capability at lower cost than can be achieved from using military-unique items.

During the Cold War, critical peer review and competition in basic and applied research was generally limited to inter-laboratory activities, and reporting of research progress and results was shrouded in a veil of secrecy in the name of national security. The requirement for this degree of secrecy has changed for both researchers and facilities. Basic research and early developmental work in technology are no longer protected in the way they once were, and secrecy is now limited more to the application of technology. As a result, some government R&D programs are opening up to the benefits of industry and academia inputs. Moreover, reliance on domestic technology transfer strengthens the industrial technology base, allowing the Air Force, for example, to depend less on foreign suppliers of subsystems or components. The end result is greater combat capability for the limited dollars available.

Increased efficiency in the development and supportability of systems is only one aspect of the evolving technology transfer policy. Another is the possible financial incentives companies stand to gain from access to knowledge, technology, skills, and materials that might otherwise be prohibitively expensive to acquire and manage. Government organizations have a stronger interest in working with industry since the passage of Stevenson-Wydler Act and related legislation have made cooperation between government and industry part of the mission requirements of science and technology staff of federal laboratories. Consequently, the Air Force has attempted to structure its technology transfer program to be consistent with national interests, its own strategic planning needs, and industry objectives.

METHODS OF TECHNOLOGY TRANSFER

Cooperation between government and industry can involve one or more of the research and development phases that transform an idea into an application. Cooperative technology transfer activities can also occur in the form of technology sharing or of cross-licensing. At the research stage, cooperative arrangements may be formed to conduct activities in basic research, experimental research, and applied/advanced development, corresponding to Department of Defense program elements 6.1, 6.2, and 6.3a, respectively. Consortia are frequently used for the early stages of research, while two-firm joint ventures are more common for the applied R&D. An example of a formal venture might be SEMATECH, an organization established for joint research activity in semi-conductor design and manufacturing. An example of an informal venture might be the Technology Forum conducted at the Great Lakes Industrial Technology Center (GLITeC) managed by Battelle for the National Aeronautics and Space Administration. The Technology Forum conducts one to two-day conferences featuring leading researchers and technology developers, both from government and industry, who have developed technology that has the potential to address specific technology interests of an individual company or an industry organization. Many other avenues exist to allow members of industrial, professional, or academic institutions to gain access to shared technology and skills.

Cooperative research and development agreements, or the CRDAs we have mentioned, are among the most popular mechanisms for technology transfer, and are the key topic of this book. They have particular strengths and weaknesses that we will explore in detail in Chapter Five. Intellectual property, indemnification, protection of proprietary material, and government licensing rights are frequently the subject of disputes between industry and government. These issues of contention, sometimes exaggerated, have emerged from past ways of doing business. Potential cooperative R&D partners who have learned to navigate their way through the issues are finding that essential legal protection and other solutions are available. The reader is referred to Appendix E for a list and explanation of the other technology transfer mechanisms available.

COOPERATIVE R&D TODAY

Early successes have encouraged greater interest in cooperative R&D. However, recent figures show a decline in the CRDA count at the Department of Energy, and at the Department of Defense. Accomplishments generally are the result of the vision and dedication demonstrated by individuals who have taken the initiative and forged working partnerships while others are still considering ways to introduce themselves as potential partners. These few individuals, largely unrecognized and unremunerated, have been pioneers in mapping out the technology transfer process. Opposition from colleagues, as well as from management, still exists; and there are limited incentives to take risks and invest capital. Technology transfer is still insignificant compared to the overall operation of a laboratory. This is understandable, given the size of federal budget cuts, the scope of program cancellations, and the reductions in personnel that have redefined and threatened jobs. This is the ongoing challenge facing those attempting to transform the mandate of technology transfer into a vibrant program of successful working relationships benefiting industry and government partners.

Community Views

In the process of conducting this study, the *Technology Exchange* research team collected and summarized the views of many people currently involved in the technology transfer process. Frank and candid discussions enabled the team to identify several themes about the current state of technology transfer. Highlights are given below.

- Word of mouth is a very important method for learning about technology transfer opportunities.
- Offices of Research and Technology Application (ORTAs) should be more proactive in their outreach efforts to industry. There is also room for state and local governments, and even educational institutions, to become more involved in technology transfer.
- CRDAs are a very useful technology transfer mechanism, as they explicitly protect the intellectual property rights of the participants. They are not governed by provisions of the Federal Acquisition Regulation (FAR) and, therefore, offer great flexibility to both industry and government.
- A serious impediment to the technology transfer process is the use of adversarial tactics by lawyers during CRDA negotiations. On the government side, adversarial behavior seems to be rooted in years of legal experience with the FAR in preserving competition. On the industry side, adversarial behavior seems to be rooted in lack of experience dealing with the government.
- Government and industry have significantly different views on marketing. Many in government believe that a dollar spent on marketing is a dollar lost on research. Industry, on the other hand, has long recognized the importance of marketing and its relationship to sales. The benefit to government is less direct, so the government has trouble justifying marketing dollars. Marketing is seen as overhead at a time when there is a push to bring down overhead. Some of the most successful government users have integrated cooperative R&D into their in-house and or funded programs so that the partner's work has a direct benefit back to the laboratory's program, justifying the marketing and integration expense.

Government and industry have significantly different views on marketing.

Government and industry have significantly different views on access to technology.

- Government and industry have significantly different views on the basis for access to technology. From the government side, it is not uncommon to hear arguments that industry should not have exclusive access to federally sponsored technology because it was paid for by all taxpayers and consequently should be in the public domain. On the industry side, arguments are made for exclusive access precisely because the know-how has already been paid for with public money, and exclusive rights will protect the investment needed to commercialize the technology.
- The technology transfer process can benefit greatly from increased support from both government and industry leaders.

From study results, a technology transfer and cooperative R&D program generally is not considered to be critically important to private companies except in areas for which unique government facilities, equipment, or test capabilities are required. Technology transfer and cooperative R&D are somewhat more important to government participants if they can forecast tangible results stemming from cooperative activity and then integrate the results into their core program or mission support infrastructure. The jury is still out on whether a dual-use technology application or a technology transfer initiative complements core activities. Within the federal defense community technology transfer is not yet considered to be integral to the strategic investment planning process. The government has barely tested the waters when it comes to integrating dual-use technology or technology transfer into a coherent long-term technology investment planning process. Rather, the general focus has been on maintaining core competencies and infrastructure as synonymous with its long-range plan. Only a few organizations have taken the first steps toward establishing a cooperative R&D program.

Technology transfer is not yet considered to be integral to the strategic investment planning process.

The next section introduces the first segment of a compendium of best practices and lessons learned in technology transfer and cooperative R&D initiatives. It has been compiled from face-to-face interviews with key leaders in successful enterprises, both public and private, who agreed to share their own organization's experiences and the knowledge gained on a non-attribution basis to assist in this study effort. The compendium offers the reader some first-hand evidence of what does and does not work in those activities that impact the success of technology transfer and cooperative R&D ventures. The following experiences emphasize the value of strategic planning. Our compendium will refer to other examples later, as appropriate.

Successful Strategy, Structure, and Focus

A few organizations have taken a clear lead in the integration and structuring of the technology transfer program into their corporate strategy and investment planning process. Characteristics common to these enterprises include a focus on having the right people participating in the business of technology transfer and on pursuing what they considered the right technologies. Another characteristic of organizations that considered themselves to be engaged in successful cooperative ventures, is the corporate recognition of "crisis," or, at the very least, a realization that geopolitical and economic changes in the world necessitate a major reexamination of their customers, their products, and their business processes.

In one success story, Sandia National Laboratory, Albuquerque, New Mexico, developed a program structure, oriented towards building "industrial partnerships" and using a team concept in which efforts were led by the product area managers. The team's leadership was technology-based, and its membership was versed in business practices. They received support through matrixed management, which resulted in a program synergy that fostered the partnership process. The concept made technology transfer everyone's job and thus ensured "the right people" in product line positions, as well as staff activities, were involved in pursuing commercial applications for its technologies.

This Department of Energy laboratory attributed its success to early recognition of an entirely changing market and top-level management support for the integration of technology transfer and cooperative R&D activities into its overall business strategy. A lesson learned here is that this type of strategic planning

Strategic planning seems to be essential to counter the effects of shrinking budgets.

seems to be essential to counter the impacts of the shrinking federal budgets, particularly in the Departments of Energy and Defense. The government R&D manager may consider making technology transfer a corporate activity to help preserve the requisite level of research activity and funding for important program efforts. A strategy is needed to overcome economy of scale problems in the federal system, and to identify and leverage essential mission support investments through the restructuring of that funding.

Another innovative program structuring concept, this one successfully implemented by Phillips Laboratory, Kirtland Air Force Base, New Mexico, was the formation of a technology applications group that was isolated from day-to-day weapon system development programmatics and empowered to pursue technology transfer initiatives. This approach offered operational flexibility for an organization in which the overall mission and mission relationships were evolving, in which specific rapidly changing technologies could be identified and pursued on a cooperative basis, and where a limited budget could be leveraged through partnership type arrangements. The interactions with industry are now helping that laboratory meet its military responsibilities by bringing in (i.e., "spin on") technology information not previously available.

Other noteworthy aspects of the Phillips Laboratory technology applications group's operations were (1) its efforts to bring "users" or systems office (i.e., product) technicians into the laboratory to facilitate effective applications for the technology, and (2) its outreach program to marry specialized technology development, test, and production capabilities in partnerships with a major Department of Defense production depot, local business, and academia.

A third example of best practices in the area of corporate strategy and program structuring for cooperative R&D comes from a leading West Coast firm in the highly competitive computer electronics industry. This organization enjoyed outstanding success with a formal technology program, which included a program vision statement, goals, metrics, and staff empowerment. Significant features of the program were (1) the active involvement of the cooperative R&D manager throughout the process, (2) a working familiarity with government bureaucracy and negotiation parameters, and (3) the emphasis on product (i.e., the technology) and solutions that work. A model was developed and offered to government organizations for structuring a technology transfer and cooperative R&D program.

The model begins with steps to establish a broad matrix of laboratory technologies and expertise available, followed by a validation step to specifically identify those considered useful by industry. At this stage, a strategic investment analysis can be used to provide a multigeneration and multipurpose road map of industry needs. Using this model, government agencies could determine where they should do research and what capabilities should be mapped for better defined planning. Other key parts of the model include finding customers (e.g., through bridging organizations), and focusing on specific products and technology (as opposed to process work, where the interaction is lacking).

Important lessons learned from the formal technology program mentioned above include the need to avoid a "clearinghouse" mentality and the benefits of working as a team in parallel, not in serial fashion. Finally, when put into place, the program should not require a large staff, and must avoid "getting in the way of the technologists."

Other people who were involved in successful technology transfer efforts offered some additional planning advice. For example, it is helpful to structure a streamlined technology transfer program on a non-adversarial basis, particularly within a culture staffed and programmed to the procurement process. It should be decentralized to the maximum extent possible, and targeted toward specific worthwhile technologies rather than homogeneously applied in a process oriented environment. Finally, the technology transfer program should be implemented incrementally, in "bite size" steps if necessary, in order to encourage effective participation, particularly where funding is austere or funding mechanisms and related processes require change.

SHOULD I CONSIDER COOPERATIVE R&D?

This introductory chapter has surveyed a lot of ground. It started with an explanation of the global situation, outlined key policy decisions, showed the macro and the microscopic view of the technology transfer budget, and discussed the state of cooperative R&D today. At this point, the reader may ask, "What's in it for me?" or "Why bother?"

Industry participants may find that government/industry cooperative R&D is right for them for the following reasons:

- CRDAs can position a firm for being "first to market." Intellectual property rights are shared with the government; however, commercial rights most often rest with firms that are willing to commit to CRDAs.
- CRDAs offer access to unique facilities—perhaps available only in government laboratories—thus allowing industry to avoid large capital outlays to purchase such equipment. Examples include large environmental test chambers, wind tunnels, and highly specialized test and range facilities.
- CRDAs offer access to experienced scientific and engineering staff, who can provide unique perspectives and expertise. Cooperative R&D with government laboratories may complement industrial core capabilities, allowing firms to try out an idea or time-phase its development before having to completely commit resources to its development. This enables firms to minimize or reduce risk.

Government participants may find that government/industry cooperative R&D is right for them for the following reasons:

- Royalties from patent licensing agreements are returned to both the inventor and the inventor's organization.
- CRDAs leverage industrial capital which may allow for achieving a "critical mass" of resources. This can help counterbalance the effects of downsizing and budget cuts at government laboratories.
- CRDAs can accelerate the R&D advances of an in-house program. Industry experience in stream-

lining product cycle times can offer cost-efficient solutions to engineering problems.

- Success in technology transfer and application of dual-use technology will ultimately lower the unit price for defense components. Per unit prices go down as the cost to develop is spread over a larger market base and as the organization moves along the learning curve. For the government program managers this means bringing systems to the fielding stage at a significantly lower cost; in fact, this may be the only way to make some products affordable.

The rest of this guidebook maps out the terrain that lies between you and the objective, Technology Exchange. It describes the obstacles you will face and the cultural gap you must bridge, as well as the best practices and lessons learned by others who have already blazed a trail to reach a viable cooperative research and development agreement.

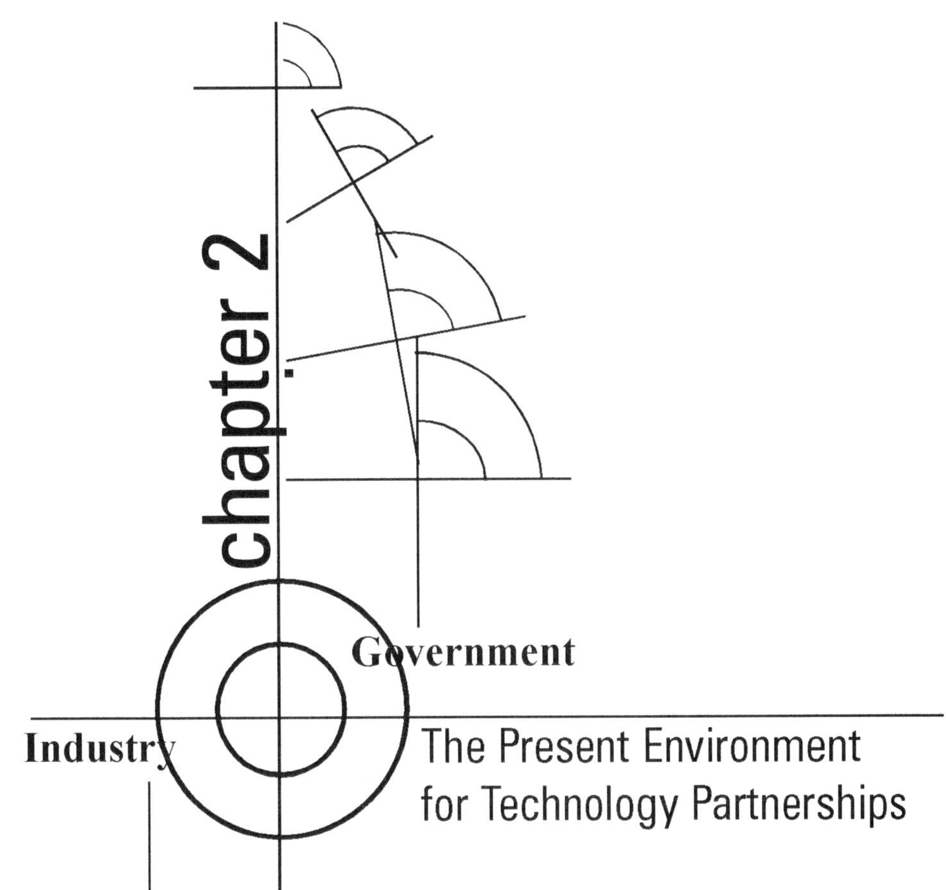

The Present Environment for Technology Partnerships

This chapter presents a survey of some environmental, cultural, and organizational variables that impact technology transfer and cooperative government and industry relations. The insights introduced here are, in many cases, based on unrefined inputs but represent accurate research results nonetheless. They are perceptions expressed by many individuals, both in the private and public sector, whom we encountered in the course of surveys, interviews, and high-level roundtable discussions. The primary goal of this chapter is to report these perceptions, not to filter them into some conclusion or recommended course of action. Chapter 3 examines many of the insights introduced here in a more structured way, showing how they impact the effectiveness of technology transfer and cooperative R&D activities. First, we provide a brief background description of historical relationships and then some honest views espoused by each community.

The Semi-Dialog—A Relatively Broad Communications Gap

[Government/industry interactions are sometimes frustrated by faulty communications. These figures may have a common subject, but they are talking past each other.]

Throughout the 40-year span of the Cold War, the federal defense establishment played the lead role in defense procurement and program oversight with great passion. Industry played the supporting role of supplying people, materials, and production facilities to the mission effort with equal fervor. The drama that played out in many cases was a confrontation over accounting methods, billing procedures, and process controls. Industry reacted negatively to what it considered intrusions into their business practices, while government officials argued that they were acting as stewards of the taxpayers' dollars. In hindsight, critics on the side of industry could claim that the government was overacting, while critics on the government side could say that industry was overreacting.

Since the end of the Cold War, leaders in government and industry have recognized the need to adapt to a different environment. That realization translates to an imperative for minimum bureaucracy, clear communications, better understanding, and more cooperative behavior between the public and private sectors.

However, cultural biases persist in this environment of change. As we mentioned in Chapter 1, one difficulty in executing technology transfer and cooperative R&D is the philosophical difference as to how taxpayer funded technology should be transferred. This difference is large enough to pose a cultural gap that must be bridged if technology transfer and cooperative R&D initiatives are to succeed. However, the effects of more than 40 years of conditioning are not easily overcome. The challenge to do so is substantial, but equally great is the reward for those both in industry and government who engage in technology transfer and cooperative R&D.

Cooperative R&D expands the field of play and interaction beyond the classic government/defense industry scope to include firms that are largely nondefense. By facilitating better communication and collaboration between the public and private sectors, cooperative R&D opens up new opportunties: industry gains access to technology that it may be able to convert to commercial application; government gains commitments that may directly support existing and future programs or procurements. Whether it is based on basic or applied research, or resides in unlicensed prototypes, much technology produced by government laboratories has yet to be applied, developed, or commercialized. However, before potential participants can embrace the vision and policies that allow technology transfer and promote cooperative R&D, several perceptions must be challenged. Government simply must recognize that industry not only desires to make a profit in their activities—profit is the driver of industrial activity. Industry, on the other hand, must dispel the notion that government can only be a sluggish bureaucracy, and learn what can be accomplished with the evolving enabling technology policy, by participating in dialog on needed process improvements and by conducting some structured inquiries on federal capabilities.

Before we can bring potential partners into the new technology transfer process, we must challenge their old perceptions of each other.

GOVERNMENT'S PERCEPTIONS ON TECHNOLOGY EXCHANGE WITH INDUSTRY

Performance, reliability, quality, and technology have always been the benchmarks for government scientists and researchers. Historically, they have been directed to spare little expense and to pay attention to the reinforcement of the national security. With the rare exception of imminent national crisis, there has been much less sensitivity to time. The scope of the mission, the customer relationships, and the technology cycle times have been entirely different from that of the commercial sector. The following paragraphs reflect views shared by government technology community members of the technology transfer and cooperative R&D program, the behavior of their industry counterparts, and the predominant barriers between the two sectors.

Technology Transfer and Cooperative R&D Programs

Technology transfer and cooperative R&D policies and program objectives have been met with skepticism in many branches of the government. There is no doubt that these policies and programs are challenged in the context of added mission requirements. Technology transfer and cooperative R&D were made additional responsibilities at a time when government programs were facing massive cutbacks. This has threatened their ability to accomplish their primary missions, and has narrowed their focus to only mission essential elements.

Many in the federal government point to the successful turnaround in technology transfer and cooperative R&D that has been achieved by the Department of Energy, the clear exception to an otherwise sporadic performance in government agencies. Its programs now are funded and well organized, in contrast to its poor position just two years ago, before the DOE integrated technology transfer and cooperative R&D into corporate planning. However, the realization that technology transfer and cooperative R&D can have an impact on mission performance is spreading beyond the Department of Energy. Technology transfer and cooperative R&D are increasingly being recognized as activities that can have a payback, and will enhance mission performance.

The leaders in cooperative R&D have demonstrated that technology transfer must be bi-directional, a win-win situation.

One perception that limits enthusiastic participation, within government, is that technology transfer and cooperative R&D follows a one-way street, meaning it represents the transfer of federally developed technology to the commercial sector, and provides little or nothing in return. The leaders in technology transfer and cooperative R&D in government today have demonstrated that technology transfer not only can be, but must be bi-directional. These pioneers and successful practitioners have proven that technology transfer and cooperative R&D, executed with the understanding and support of the corporate leadership, can be successful if it is a win-win situation. They cite the phenomenal investment leverage that can be gained through cooperative R&D. In some cases, industry investments have doubled the value of the organization's original investment. They also reference the long-term economic and national security benefits of introducing otherwise latent technology into the commercial sector.

The government laboratories are not interested in commercializing or developing products for a large market. Their expertise ends where consistency, high volume, pricing sensitivity, quality, and economies of scale become issues of importance. Technology transfer is an enabling process that fosters the infusion of technology and processes that lead to future commercial development of products and systems that, in many cases, will be available to the U.S. government.

Many government technologists are optimistic about the benefits of cooperative R&D. It allows them to invent and create without overwhelming commercial burdens; it is insulated from fickle market forces and preserves otherwise threatened programs. In fact, during the benchmarking visits to several government laboratories, we frequently heard the opinion expressed that the best chance for survival may be through cooperative R&D.

Technology transfer and cooperative R&D programs are young. Consequently, they may offer a return on investment that is too intangible for many in government, but they do offer the federal research laboratories an opportunity to reinvent themselves. This is critical as budgets shrink and the need for sophisticated weapons R&D programs is subjected to increased scrutiny.

Technology transfer, and the joint strategic planning that supports it, may provide the impetus for a laboratory to redefine its mission and preserve its core competencies.

Industry Behavior

In the course of our study, we found that some in government are skeptical of the merits of cooperative R&D. They question the kind of R&D commitment they see in the private sector, and find little that would encourage government to undertake cooperative efforts. These individuals resent the notion of financial windfalls to be reaped by industry based on government funded research. Conditioned by the government culture, they argue that industry is too driven by the profit motive and assert that effectiveness is sacrificed in the name of efficiency. Some assert that industry jumps from one point of commercial interest to another, seeking out targets of opportunity and measuring the timeliness of cooperative ventures more than ensuring the quality of research being conducted.

INDUSTRY'S PERCEPTIONS ON TECHNOLOGY EXCHANGE WITH GOVERNMENT

Any discussion of technology transfer and cooperative R&D must address the attraction between these two disparate cultures, and the prospects for them to come together in joint R&D and other cooperative programs. In the absence of compelling forces, it is unlikely the two sectors would cooperate. However, strong international competition, a less dominant domestic economy, and the increasing costs of R&D at a time where a short-term, "bottom line" mentality dominates the defense acquisition and R&D decision-making process, all make cooperative R&D and technology transfer an attractive, cost-effective option. Not surprisingly, the Department of Defense's long-term planning, technology research, and weapons development activities are threatened by similar pressures.

Technology Transfer and Cooperative R&D Programs

Industry sees many attractive benefits in cooperative R&D. Cooperative ventures provide the opportunity to receive technical information and expertise not available in the private sector. A significant benefit is access to government R&D expertise, unique facilities, and funding (although under CRDAs there is no direct funding from government).

Government should have a clearer understanding of the goals and motivations driving industry to technology transfer and cooperative R&D. While rallying forces to build the technology base and enhance the national security may be effective as a rhetorical device, it generally does not facilitate increases in revenue. Private corporations are in business to generate profits for their shareholders by creating and marketing revenue-generating products and services. Business is constantly looking for ways to reduce the internal research and development costs of future products and services, while reducing time to market.

As the international marketplace becomes more competitive, industry requires designs or models of commercially viable products that have the potential to be quickly introduced into markets. In a technology transfer and cooperative R&D program, technology from the federal laboratories would be used to both produce profits and sustain long-term market share. Government and business are now facing similar operating constraints: limited resources, accelerated technology cycle times, and downsized budgets and staffs. However, industry, mindful of years of experience operating under the federal acquisition requirements, is generally skeptical of government's ability to move toward truly cooperative R&D behavior.

Cooperative R&D programs would enable the private sector to develop technology from federal laboratories to produce profits and sustain long-term market share.

Industry is especially concerned with timely results. The management and staff of commercial organizations must continuously monitor themselves to detect and modify inefficient behavior and administrative practices in order to remain efficient and be competitive. They do not perceive this type of behavior in government.

Industry is interested in acquiring and protecting intellectual property rights. In cooperative ventures, one industry concern is that the bureaucratic mishandling by the government may result in release of proprietary information and trade secrets. Furthermore, even though the CRDA is perhaps the most user friendly mechanism the government laboratories have, many in industry perceive the entire CRDA process as becoming too legalistic and contractual.

Government Behavior

Industry views government as too bureaucratic, requiring specifications and review/approval times that overwhelm and frustrate those who attempt to negotiate a CRDA. Industry also views government as overly restricted by procedure and tradition, and still unwilling to accept the notion that the federal laboratories may require reform, or that the post-Cold War reality may dictate consolidation and downsizing. Industry also feels that government does not understand non-government needs. While industry is struggling to capture international market share, it feels that government neglects commercial concerns regarding cost and timeliness.

A particularly striking perception in industry is that the federal government and private business have developed completely different business operations, each bolstered by views that now act as roadblocks to cooperation. One such roadblock is the government's inability to fully accept the concept of profit. For industry it is an absolute—for government, it is something to be controlled, monitored, and in some cases challenged. Industry often complains that the government simply does not understand the role that profit generally plays in private business, and specifically the linkage between profit, product differentiation, and corporate legacy on the one hand, and intellectual property, exclusive license agreements, and proprietary rights on the other hand. In industry, having the latter is essential to maintaining the former.

Results of our study indicate that industry perceives government laboratory, legal, and agency staff as combative, rather than cooperative, in the agreement negotiation process. The reason postulated for this behavior is that many of these individuals are unable to view CRDAs outside of the federal acquisition requirements. Lacking clear guidance, they are unwilling or unable to influence the organizational and cultural factors vital to managing technology transfer and cooperative R&D. Industry is concerned that a change in behavior within the federal laboratory system is thwarted by a number of major obstacles, including a lack of awareness of business drivers; the government's perception of a cash-driven, short-term perspective in industry; issues of control, influence, and objectivity of the laboratory; and confusion of the procurement process with the cooperative agreement process. Industry is not going to be impressed by incremental change; a total culture change is needed.

Our study indicates that technology transfer and cooperative R&D policies are inadequately funded and lack clear performance criteria. To industry, this suggests a lack of commitment on the part of the federal leadership. While technology transfer legislation was conceptualized during the end of the Cold War, the public law was not implemented until many federal agencies were facing drastic cutbacks in budgets and operations.

BARRIERS TO COMMUNICATIONS

Many in government believe that the differences noted above constitute a cultural gap that is simply too large to bridge. Furthermore, the cultural differences exist in many forms, varying from organization to organization and even within organizations, resulting in a complex set of cross-cultural conflicts. Overcoming the internal cultural conflicts of government and industry laboratories is the key to fostering overall organizational interest and involvement.

Many in industry believe that the lack of familiarity of government officials with business and finance, profit and loss, is so profound that it represents a significant barrier to successful communication. Although significant acquisition reform initiatives are being considered within the government, this perception is exacerbated by a cost accounting system driven by provisions of the Federal Acquisition Regulation (FAR) that nondefense companies cannot afford to adopt. Furthermore, industry is generally unsure as to government's commitment to cooperative R&D. This lack of certainty cools industry's commitment to technology transfer as well. However, the Internet and World Wide Web have made it easier to access information about the goals and objectives, short term and long term, of a potential R&D partner, thus improving some on mutual understanding.

Acquisition reforms have come a long way to reducing—but not yet eliminating—the barriers to communication between government and industry. Each entity has its particular driving force: in the military it is mission, in industry it is profit. Prospects for mutual understanding have improved with the emergence of tools like the Internet and the World Wide Web, which make it easier to access information about the goals and objectives of a potential R&D partner.

Many government representatives agree with the industry perception that federal bureaucracy and a lack of business expertise thwarts attempts at effective communication, particularly when addressing market-driven problems. Government technologists admit that they lack sufficient background in product manufacturing, marketing, and distribution. This inexperience often renders their designs insensitive to cost, unrealistic in terms of development schedules, and incompatible with market niches.

In summary, significant differences in the perspectives of government and industry can and do impede progress in cooperative ventures. As both sides realize that they need each other's perspectives and combined resources to survive global competition and effectively manage shrinking resources, their goals and procedures will change toward becoming more and more cooperative. Good communications can be a key to identifying, understanding, and overcoming culturally derived barriers to this process.

chapter 3

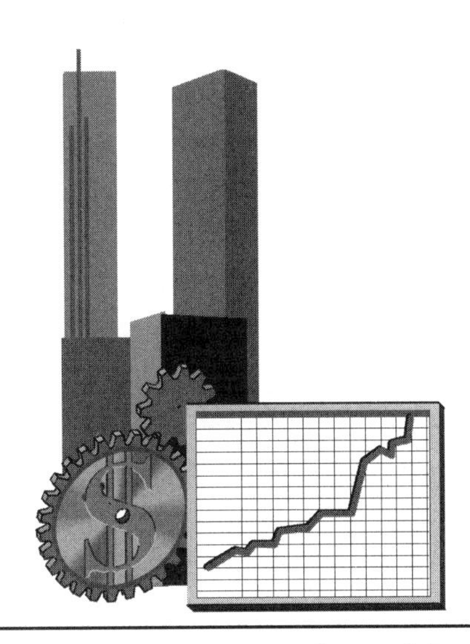

Cross-Cultural Considerations and Imperatives

All the labs are different. Industry is different. We're trying to do something which is difficult even between two human beings. I was involved in a small company once: I found out how difficult it was to cause technology transfer between one physicist and one engineer. You're now talking about trying to cause it to take place between institutions that are large and have their own characteristics and dealing with industrial organizations, each one is unique and has its own characteristics.

—Alvin W. Trivelpiece, Director of the Oak Ridge National Laboratory in Testimony to the U.S. Congress

Public and private sector technologists, managers, and administrators all face major barriers to effective communications that can impede successful technology exchange. Communications break-downs may be self-sustaining in a single organization, but in technology transfer activities between organizations they can be symptomatic of severe cultural clash. During the Cold War period of more than forty years, two distinct cultures developed in industry and government, resulting in, at times, an adversarial relationship. Each party has established a rigid set of behavioral norms, and different measures of performance and effectiveness. These distinct cultures view, create, and nourish barriers that must be overcome for cooperative ventures to be successful.

People with careers in industry, government, and academia have markedly different perspectives of the world—perspectives that clash in teaming arrangements.

The rules, norms, beliefs, values, and philosophies that are shared among members of a group define that group's culture and create a powerful force that guides individual behavior. Culture provides meaning, direction, and motivation to an organization, and acts as a lens through which the organization as a whole and its individual members view the outside world. The people whose careers are spent in industry, government, and even academia all have markedly different perspectives of the world, and these perspectives often clash in teaming arrangements like CRDAs. This has certainly been the case in the evolving relationships between the various science and technology (S&T) communities, engaged in research, development, testing, applied engineering, production, and support activities.

For those who are patient and persistent in overcoming the cultural differences—and understanding the strengths and weaknesses of the unique perspectives—the rewards can be impressive.

Many technology managers struggle to identify and contact suitable partners for cooperative R&D. Once they have succeeded in communicating their needs and goals, they believe that the agreement will take care of itself. Inevitably, much negotiating work lies ahead, and common misperceptions about cooperative R&D often get in the way. Reconciling the cross-cultural differences is fundamental to successful technology transfer and cooperative R&D. A basic understanding of the unique perspectives which government, business, and academia bring to negotiating and the other aspects of the cooperative R&D process will greatly increase the prospects of a positive technology transfer or cooperative R&D experience.

In the survey conducted by the *Technology Exchange* research team, the majority of respondents from both industry and government reported that the area of communication needed significant reform. The literature on cross-cultural communications within a technical community suggests, in fact, that language used by these two sectors (namely military agencies and corporate R&D staff) is frequently only a "semi-dialog,"[1] as we illustrated in the graphic at the beginning of Chapter Two. However, there are instances when the gap narrows somewhat to indirect communications, if not com-

The Dialog—A Relatively Narrow Communications Gap

pletely to highly desirable direct communications. Organizations in the Department of Energy, for example, are undergoing a total culture change, making technology transfer an integrated corporate activity. Other government agencies are showing signs of slow progress in this area as well, pledging increased leadership support for technology transfer programs, and making commitments to pursue opportunities for successful technology partnerships every day.

Contrasting cultural norms are particularly evident in the five areas discussed in this chapter. They are the most basic, deep-seated, and therefore perhaps most challenging obstacles to direct communication within the public and private S&T communities. As we examine them we will pay particular attention to the ways these obstacles interrelate, how they affect the technology transfer process, and how much the different viewpoints can affect the road selected to the ultimate destination: technology partnerships.

MISSION CRITICAL SUPPORT AND CORE COMPETENCIES

The Concept Of "Mission"

Dedication to mission is an important factor for success in the defense community.

One of the starkest contrasts in culture between the government and the private sector that impacts the effectiveness of cooperative research and development stems from the defense worker's orientation towards the "primary" military mission. The government employee's traditional views towards work, technical performance standards, and job rating or success criteria, for example, are often alien to the industrial marketplace competitor, or to those in academia. The various kinds and amounts of support required to meet core research, development, test, production, operation, and sustainment capabilities contribute to the complex set of cross-cultural conflicts. The differences include attitudes about and approaches to methods, processes, priorities, structure, quality standards, means of control, accountability, and decision-making. Dedication to mission is an important driving factor for success within the defense community, matching, as we will see later, the role of profit for private enterprise in bringing a product to market.

The traditional view is that the primary mission for the federal defense employee is to develop, produce, enhance, and support the military systems that provide a war fighting capability for the U.S. that is second to none. There is no question that this charter for the defense establishment and this vision for military power are completely valid. The conflicts arise when a perceived secondary mission is added to the goals and objectives of a federal workforce already affected by budget cuts and reductions in staff.

Specifically, technology transfer and cooperative R&D activities were envisioned and incorporated into policy in order to realize the maximum benefit from U.S. national technological resources. The rewards, theoretically, are two-fold. First, technology transfer and cooperative R&D are considered to be a way to stimulate the domestic economy in an increasingly competitive international marketplace in order to retain American strength and meet our federal budget responsibilities. Second, they act as a force multiplier in the form of an expanded industrial base for the defense conversion effort, thereby providing, in the long term, better supportability for military systems. The policy behind these activities, as it has evolved through the encouragement of various cooperative technology partnerships, defense conversion, and dual-use initiatives, is good only if it is understood and broadly supported by those who must play a major part in implementing it.

Our study showed that, by and large, technology transfer continues to be seen as a one-way flow from the public sector to the private sector, and not for its potential as a two-way exchange of technology. Recently, technology transfer has been touted as an important mission element by the Air Force, the principal subject of our research. This is certainly progress and an important step; however, until it is recognized and incorporated as an integral part of the organization's technology investment strategy, it will remain a secondary mission, and technology transfer successes will continue to follow more from individual efforts than from a natural evolution to an institutionalized, well understood, and purposeful business practice.

Concerns About Core Competencies

A concept related to mission, at least from a military leader's perspective, is the control or preservation of technological core competencies in order to satisfy system support requirements throughout the spectrum of conflict, from peacetime to hostilities. The need for access to core competencies extends from the laboratory R&D environment into the product life cycle. An important concern among many defense technologists is the need to maintain certain core competencies that may be "world class" or considered "one-of-a-kind" at a laboratory or technology center in order to continue providing a basic defense capability and the capacity to surge when needed.

Exporting core competencies to industry is perceived in some cases as a dilution of this capability, potentially affecting the defense employee's ability to support the primary mission. When linked to defense downsizing and base closure activity, it is also seen as a direct job threat. The export of core competencies to industry, with its very different vision and mission, seems to have dubious value to many defense workers for the following reason: there is a fear that the results of these activities will inevitably leave the military without some R&D capability which is viewed as essential and irreplaceable, and which the private sector could not or may not choose to maintain.

The concern is that core competencies will be lost as a result of technology transfer to the private sector.

In industry, the concepts of mission and core competencies take a different form. Here, concerns center around marketability decisions. The focus of industry is on making decisions about in-sourcing versus outsourcing for resources; determining acceptable levels of technical risk; evaluating comparative costs; weighing competitive advantages; looking at future growth; and acting to meet long-range as well as short-range goals. This leads to the next major barrier to communications, which had and continues to have a profound impact on the success of technology transfer and cooperative R&D activities. That subject, of course, is profits and the funding that makes things happen in a free and competitive enterprise system.

PROFITS AND COOPERATIVE R&D PROGRAM FUNDING

Where mission and the maintenance of the public trust drive the federal establishment, it is the concept of profit that makes the business world go around. It is what motivates the risk taker, stimulates excellence, and filters out inefficiency. While the term "profit" is not alien to the government worker, it is viewed by nearly all of those involved in fiscal matters within the federal business process as something to minimize rather than to actively pursue. A cost conscious, excess profit, "watchdog" mentality is the dominant cultural factor that the private sector reacts to in its adversarial relationship with the public sector.

Corporations strive to achieve technology superiority; they are hostile to the notion of sharing new discoveries.

Corporations constantly strive to acquire solid proprietary holds and technology superiority which will make them more competitive. Few companies are interested in paying for or licensing a technology they cannot exclusively control. Corporations are hostile to the notion of sharing intellectual property or new discoveries. The insistence on non-exclusive licensing by government has in some cases resulted in no one person or organization acquiring the technology; consequently, no technology transfer was achieved, which would appear to be counter to the intent of technology transfer legislation. Corporations need to be aware of the fact that the partners in CRDAs get the first right to exclusive control of intellectual property.

Many large commercial firms that have worked with federal agencies and laboratories have expressed mixed emotions about their experience. While there is generally a high level of enthusiasm, the way in which commercial firms are viewed and treated by their government partner varies with agency and laboratory. The government in many cases is seeking to insert technology into the commercial domain for development into products which it can ultimately purchase in commercial lots. However, the government is often uncomfortable with commercial scales of profitability. If the government is inclined to make purchases at prices consistent with commercial markets, then it will also have to recognize the scales of profitability required in most commercial firms.

Another cultural difference that widens the gap between the two sectors is the funding process used by government when it pursues an activity such as cooperative R&D. This is a substantial barrier, stemming from differing cultural mindsets towards profit and exacerbated by a business process that is heavily weighted towards cost accountability. The type, amount, and timing of funding for activities is an important aspect of business operations in the private sector. Funding represents a commitment to an activity. At a very basic level, it can be tied to a spending plan and the outlays can be tracked against accomplishments. On a more complex level, it can be tied to the projected profitability of the activity; funding can be increased, decreased, or cut, according to evaluations of results obtained. This cultural complex around funding does not fit well in government/industry cooperative R&D, where industry is generally asked to provide money, while government agencies make their contribution, as guided by Congress, in the form of time, materials, equipment, and expertise. Potential partners who are uncomfortable with this one-sided funding arrangement should look into cooperative agreements other than CRDAs, where cost sharing is permitted (see Appendix E).

A change in budget orientation may signal change in culture and perhaps mission. For example, it is estimated that government organizations devote less than 3 percent of their research budgets to technology transfer and cooperative programs. Government agencies that do not specifically budget technology transfer activities are likely to support them by assigning technology transfer responsibilities as an additional administrative duty. To facilitate change, the Clinton Administration has expressed a desire that all federal organizations commit more of their R&D to dual-use, technology transfer, or cooperative R&D activities. Percentages vary from agency to agency.

These and other cultural clashes are explored further as organizational biases in the following paragraphs. After a review of the government side, we examine the industry side, distinguishing between (1) medium and large size companies, (2) small businesses, (3) defense contractors, and (4) academia. As we will see in the next section, it is the intellectual property issue rather than other issues (e.g., the number of staff and employees) that is most important for cooperative R&D arrangements.

Organizational Biases on the Government Side

All federal agencies that engage in research and development are mandated by Congress and Executive Order to engage in technology transfer and cooperative research and development with state and local governments, as well as with industry and academia. In conjunction with federal and agency mandates, they have the responsibility to transfer unclassified technology to the nation's industrial base. Each branch and agency of the U.S. government has unique aims, motivations and research interests which reflect the charter of the organization. Mission requirements and research needs establish the parameters for the type of cooperative partner they are inclined to work with.

The enabling technology transfer legislation has been in place for several years, but its implementation has been overshadowed by post Cold War downsizing. Technology transfer has been a hot political issue, with officials keen to publicly proclaim its potential. Implementing technology transfer has proven a different challenge altogether. The government acknowledges it has a great deal to gain from technology exchange activities; the lack of understanding and support on the part of mid- and upper level managers can be traced to short-term or unclear benefits, a vague selection strategy, and limited financing options.

Declining budgets, resources, and staff levels make it difficult to support and foster technology transfer. Internal resistance, lack of funds, and unsupportive mid-level management challenge those attempting to facilitate cooperative R&D. However, government researchers are increasingly taking advantage of the opportunity to work with talented and enthusiastic industry partners or develop partnerships with those few organizations which have embraced technology transfer.

Cooperative R&D opens the door to different perspectives, while also creating new avenues and sources of funding. As R&D program outlays diminish and programs are canceled for lack of funds, cooperative R&D is increasingly seen as an opportunity to leverage federal investments in basic and applied research. The need to leverage scarce resources acknowledges a simple reality: the Federal Government, and the Department of Defense specifically, cannot afford to finance its own industrial base any longer while the demand for new and upgraded weapons platforms remains stable. Philosophically, the Department of Defense views technology transfer as an opportunity to transfer technology and techniques to the industrial base, from which it hopes—in the long term—to procure military hardware in the future.

The Department of Defense views technology transfer as an opportunity to ensure that the private sector has the technology and techniques needed to respond to future procurements.

The government has been quick to admit that to maintain core competencies in this rapidly changing world, working with industry is not only welcome, but critical. Cooperative R&D and technology transfer programs allow world class, market tested commercial technology and techniques to be brought to bear and applied to government research programs at low cost to the federal government. This is commonly known as "spin on" application and is a characteristic of technology transfer as a two-way exchange.

Organizational Biases on the Industry Side

Whether we are discussing a defense contractor or a fledgling company that has never worked with the government, the bottom line is universal: make money for the company and the shareholder. Both ventures generate revenue from the sale of products and services. Unique marketing requirements and specialized product development shape how corporate cultures approach technology transfer and cooperative R&D. The difference in motivations and requirements mandates addressing these highly specialized industrial cultures individually.

Technology transfer and cooperative R&D legislation was intended to benefit small and medium size companies. This was borne out of the fundamental recognition that the greatest potential for job, business, and national economic growth exists with these companies. (The actual language of the legislation cites "special consideration" for the small business, which is considered to be up to 500 employees.) It is difficult to accurately measure the success rate across small, medium, and large companies involved in cooperative activities, because no one has carefully tracked the percentage of small versus large. The majority of industry success stories in cooperative R&D and technology transfer appear to be come from large companies. The likely reason for this is that the large companies will talk about successes; small companies generally do not talk about their experiences because it is perceived that it hurts competitiveness to talk "out of school."

It is difficult to assess the success rate because no one has collected the data. Most of what we know comes from firms willing to share their experience.

Medium and Large Size Corporations

Most corporations that engage in research and development face pressures from competition and increasing costs in research, product development, and marketing. While cost saving strategies are similar to those found in government, small business and academia, there are differences. The priorities for large commercial firms (i.e., those that do a majority of their business in the commercial market) are as follows:

- Leverage internal research efforts
- Use technology transfer and cooperative R&D to develop new applications for existing technology
- Use federal facilities and equipment not available elsewhere, or that is prohibitive in cost to develop internally.

The criterion for success is ultimately whether or not the company makes money. This is generally accomplished through creating and marketing products which generate revenue, or by reducing costs. Most firms view technology transfer and cooperative R&D as an opportunity to leverage their internal technology base at minimum cost.

Most firms strive to sell products that are superior in quality, lower in price, or covered by better service than the competition offers. Large firms often are in a better position to do this. In many cases they have proprietary technology, which by definition cannot be utilized by another firm without licensing, that provides the competitive advantage. Technology transfer and cooperative R&D activities can give rise to considerable intellectual property and licensing complications. This motivates many large firms to opt for long-term, basic, or applied research that will not result in a technology development that might be harnessed by another firm.

Large commercial firms increasingly are becoming enthusiastic in working with the federal government. Many corporations have found their cooperative research and development agreement (CRDA) experience to be positive. They believe, however, that it can take several years before financial rewards will result from these partnerships.

Small Businesses

Many small businesses lack the resources necessary for a long-term commitment to pursue basic or applied research; they look more for a pay off in the short term. They generally pursue technology transfer in the hope of acquiring a technology, process, or technique, around which the company will be able to build a product line or business venture. Small companies have demonstrated considerable risk taking and entrepreneurship in their efforts with government. However, intellectual property and licensing disagreements often have stymied an otherwise energized process.

Many companies gamble their future in the hope the technology they develop and market will pay off. As many have learned, the successful acquisition of a technology is only a fraction of the cost. Activities associated with development, design, manufacturing, marketing, distribution, and sales represent the majority of the cost of new technologies, and generally require the infusion of capital, frequently from venture capitalists. However, many venture capitalists have little to do with start-up companies, and show interest in small companies generally only after the first $5 million or more in sales. These venture capitalists apply very strict performance standards to small companies, and the intellectual property and licensing restrictions in technology transfer and cooperative R&D may not meet these tight requirements. In several instances, small companies have become involved in technology transfer and pursued ownership of a technology under the belief that they would have exclusive international and domestic rights, ultimately to find one or the other was not available. Venture capitalists are unlikely to capitalize a firm without exclusive rights to the intellectual property and technology. This vividly illustrates the relationship between profits and the willingness to finance risk on the one hand, and intellectual property and the technology product on the other.

The adoption of technology is only one reason for examining the usefulness of technology transfer and cooperative R&D. Small business are also attracted to opportunities that grant access to expertise and state-of-the-art manufacturing, materials, and equipment. This is where some small businesses expect to find the greatest rewards.

Defense and Federal Government Contractors

As defense procurement budgets decrease, defense contractors are consolidating or are being acquired by other firms. Those that survive are diversifying their core business functions to include a greater percentage of commercial products and services. They have found that old defense competencies—innovatively applied to new situations—can result in profitable new commercial markets. One excellent example is the conversion of sophisticated radar equipment for use in tracking rainforest clear cutting in South America.

The defense industry maintains a perspective measured in decades, rather than months. Multi-billion dollar, decade-long programs are not uncommon. This cultural reality has tended to insulate the defense industry from periodic Defense Department, Presidential, and Congressional initiatives that claim to change "business as usual." Some defense contractors are skeptical of technology transfer and cooperative R&D. They wonder if technology transfer and cooperative R&D is a fad; they suspect that the claims made by federal researchers are oversold; and they question the merit of investing resources into an activity that may be suspended or deferred with the turn of an Administration or Congress.

These considerations have not kept the defense industry from taking full advantage of technology transfer and cooperative R&D as a "target of opportunity." Because of uncertainties about the duration of the program, defense companies view cooperative R&D as an interim technology resource. In fact, the defense industry is proving agile and competitive in cooperative R&D and technology transfer. Like many other companies in private industry, defense contractors are motivated by the opportunity to leverage their technology and knowledge base through long-term cooperative R&D activities.

Technology transfer and cooperative research and development are not guided or constricted by federal acquisition requirements. One of the most important opportunities afforded the defense contractor by these cooperative programs is an opportunity to get to know the customer in a non-adversarial, win-win environment.

CRDAs offer both government and industry a non-adversarial, win-win environment.

Academia

Academia enters into cooperative R&D relationships with the government from the same side as industry. However, academia does not share much of the culture of industry. Unless the work is conducted on behalf of a sponsor, universities traditionally tend to focus on basic research, and generally are not concerned with product oriented research or development. However, even this cultural reality is changing as academic researchers in some fields (biotechnology, medicine) are joining the rush to patent and commercialize their technological breakthroughs. University operating and research budgets are experiencing cutbacks as external funding from industry, and support from the state and federal government declines. Universities view cooperative R&D as well-leveraged, low-cost opportunities to share talent, knowledge, and equipment.

THE COMPETITIVE EDGE—INTELLECTUAL PROPERTY AND TECHNOLOGY CYCLE TIMES

We have shown how mission and profit are important, if not definitive cultural factors, providing the basic energy and motivation for two very different communities to conduct technology transfer and cooperative R&D activities. Improving our understanding of these very different mission support and fiscal perspectives is the only a first step. Next, we focus on two key imperatives for working cooperatively and successfully within our competitive market system. They are intellectual property and technology cycle times. Understanding them and their essential role in the cooperative R&D process is the next step in our program design and execution.

The federal technology transfer and cooperative R&D programs are an integral part of the U.S. economic strategy to strengthen our global competitiveness, and, in so doing, to enhance our weapon system support capabilities for a strong national defense. The topics of intellectual property and technology cycle times collectively define "the competitive edge" upon which business succeeds. Both elicit much differing views and behavior from within the public sector, and, therefore, become obstacles in developing partnerships between government and industry.

Intellectual Property

In the execution of cooperative agreements, one great area of contention is intellectual property. The overarching issues which impact cross-cultural relations will be discussed here. Specific legal issues which affect development of cooperative research and development agreements and other technology transfer agreements will be discussed in a subsequent chapter.

While substantial progress has been made in identifying potential intellectual property disputes, only a portion of the issues have been resolved to the satisfaction of industry. One point of contention which prohibited many from entering cooperative R&D agreements was the fear of release, for example through the Freedom of Information Act, of proprietary information generated or contributed in the course of cooperative experiences. The perception in industry is that this issue has been largely resolved. As amended by the 1989 Technology Transfer Act, proprietary information developed in the course of cooperative research and development is no longer subject to the release requirements of the Freedom of Information Act for 5 years. A period of 5 years may be long enough for industry in some areas, but not long enough in others, such as protecting dissemination of the formulas used by beverage companies in their soft drinks.

The National Technology Transfer and Advancement Act of 1995 provided a directive to the government to "not publicly disclose trade secrets or financial information that is privileged or confidential within the meaning of section 522(b) (4) of Title 5, United States Code, or which would be considered as such if it had been obtained from a non-Federal party." The act also expanded the options that a collaborating party has in the granting of patent licenses or assignments. Where earlier the legal presumption or default position was nonexclusive license, the collaborating party now may choose an exclusive license for a pre-negotiated field of use.

The law remains sensitive to an attempt by industry to pursue a license just to protect against someone else using the technology. The government retains the right to require the collaborating party to grant to a responsible applicant a nonexclusive, partially exclusive, or exclusive license to use the invention in the applicant's licensed field of use, or to grant such a license itself if the collaborating party fails to do so. The government may exercise its right only in exceptional circumstances and following a specified determination that is subject to administrative appeal and judicial review.

A license holder must take appropriate steps to commercialize a technology in a timely manner; otherwise the license may be revoked.

In order to understand why intellectual property is a wedge in industry and government relations, it is necessary to review some of the underpinnings of the debate. Philosophically, private companies typically believe that because the technology which resides in the laboratories is taxpayer funded, they should be able to access it for free. Conversely, many in government argue that since government technology is taxpayer funded, *all* should share in its benefits, and no one company should be permitted exclusive access to it. The middle ground between these two divergent perspectives is that the government will transfer and license technology, but generally not exclusively. And, as industry makes clear, without exclusive rights domestically and internationally, it may not be worth investing in or co-developing with government researchers.

Government, by nature, is not held accountable in the same way industry is. In many cases, industry is required to devote funding to cooperative R&D projects, for example the co-development of technology, which in some cases results in technology suitable for market. Government will seek to control inventions that result from its own intellectual property, and will agree, in the negotiation phase, to a fair share of royalties on inventions that are jointly developed. Government will not seek royalties on an invention that belongs to the partner. Private companies do not always understand these distinctions and argue that if they are going to be required to pay licensing rights for technology they helped pay for, their profitability will be reduced. Many government agencies do not accept this argument. Government culture takes the view that an invention by the partner belongs to the partner, an invention by the government belongs to the government, but can be licensed to the partner. The question of where the money came from to develop the invention is not addressed.

Part of the problem is that this is a mix-up of the patent license agreement and the cooperative research and development agreement process, which contributes to the communication gap between industry and government. Private companies argue that they are being asked to carry the burden of financial risk. With increased probability of financial reward, this is often acceptable. With reduced likelihood of financial reward, many in industry will continue to be reluctant to enter into cooperative agreements. Government counters with the argument that because most CRDAs have an uncertain outcome, it is not logical to debate or negotiate patent or licensing issues at the time of initiating the agreement. Further, government agencies are interested in leveraging their programs through patent licensing. Royalties from licenses go to certain narrowly specified activities, including patent expenses, technol-

In the private sector, acceptability of risk is tied to probability of financial reward. If the likelihood of financial reward is low, there will be reluctance to enter into cooperative agreements.

ogy transfer programs, training, and awards. Reimbursements from CRDAs go back to the program.

Technology Cycle Times

The competitive edge is achieved largely by the producer who gets to market first and carves out market share. Market share will often depend on the quality and technical superiority of the product. The speed in which this is done—the technology cycle time—is vital to gaining and sustaining market share. Technology cycle time can be defined as the time that it takes to research, develop, prototype, manufacture, and test market the product. Sometimes the cycle time does not include market testing. For the purpose of this discussion we will include it, arguing that the viability of a product is not proven until it has been market tested.

Technology cycle times have been a defining issue for private companies—the key to their success or failure—for a long time. Rapid advances in information technology and international competition have elevated the pressure to unprecedented levels. A company strives for quick cycle times in order to introduce superior products ahead of competitors in the most cost-effective fashion. In many cases now, technology cycle times are measured in months, rather than years. Ignoring cycle times can allow competition and, in the case of weapons systems, potential enemies to gain the advantage.

Quick cycle times are crucial for a competitive edge. In the private sector, technology cycle times often are measured in months, not years.

Government is seen as having little ability to contract their cycle times. In the past, the review and approval time for CRDAs might take as long as 18 months. However, the time that it takes from initial discussion to negotiation, closure, and signature of CRDAs has dropped dramatically over the past year. Government has improved the process through standardized CRDAs and decentralizing much of signature authority. In most cases, the CRDA process now takes between 30 days and 6 months. One large impediment to the process exists with government lawyers who are trained in procurement contract negotiations and feel it a responsibility to say why a technology transfer agreement should not be entered into, instead of how to remove impediments to the process. Where corporations use lawyers to review agreements in most cases, the government tends to rely on them to approve the activities themselves. While there are individuals who are outstanding leaders in the technology transfer community, a number of government lawyers still act in the role of government procurement officer, using federal acquisition requirements as their lodestone out of habit, if nothing else.

In addition to the competitive imperatives stemming from intellectual property and technology cycle times, there is another important cultural contrast between the government and the private sectors that must be understood and factored into the planning and implementation of technology transfer and cooperative R&D activity. In the next section, we will see how a more focused view towards the technology product itself can lead to fulfillment of cooperative investment objectives, with both internal and external customers, and to a more successful overall business investment strategy.

FOCUS ON PRODUCTS VERSUS PROCESSES

Another significant cultural barrier impacting the effectiveness of cooperative R&D stems from differences in product versus process orientation of the two communities. A dominant perception from our study participants, both public and private, was that the government, particularly the military, was overly focused on the technology transfer process at the expense of the technology product with which it works. To explore this important issue further, we should first distinguish what we mean by product, since the laboratories are set up to conduct R&D, not develop commercial products. We are referring instead to specific technologies (i.e., science if you will) as the product of a laboratory environment and not restricting our definition to that of an end item application for the marketplace. The message here is about priority and focus, not choice between these two essential components of any vibrant and productive cooperative R&D program.

Technology assessment is a crucial activity missing from several federal agencies and laboratories engaged in technology transfer. The lack of an early and comprehensive examination of what an organization does best can hinder development of efficient and effective relationships, and even cripple the structure and control of cooperative R&D activities. The government does not fund laboratories to commercialize products, so they are not conditioned to assess potential marketability of a technology or its applicability to a market need. In many cases, then, the government looks at cooperative R&D as a process, and misses the link to product. This focus on process has, in effect, preserved centralized control over resources and investment decisionmaking. Technology transfer can have a low tech or high tech focus, and the government can benefit from the expertise of industry in reading the market and adjusting to market sensitivities. The effect would be sharing resources and decentralizing decisions to cognizant and accountable business authorities.

Federal laboratories do not engage in marketable product development; there is no emphasis on identifying and nurturing the technology with the most promising commercial utility.

In our study, we saw the inclination to establish a staff or ad hoc group, empowered with a process development charter in technology transfer, to implement a program that would fit within the established organization and into the overall technology business process. This is a good thing if the right people are participating, if critical strategic planning is done, and if the work being accomplished results in good investment solutions for "spin on" for government or "spin off" for commercial applications of the technology. On the other hand, if this staffing activity, program structuring, and process development precedes (or is done in isolation from) an overall corporate examination of technology needs and exploitable capabilities, then the effort may be too costly and time-consuming to be worthwhile to anyone. Moreover, the result will likely not be part of an integrated investment strategy developed by those responsible for the technology.

In one program we studied, considerable team effort was expended planning and defining a "master process" to enable the decentralized development and execution of technology transfer business investment plans at the laboratories and other assigned technology centers. The loosely defined process was designed to provide an enabling framework for the identification and transfer of appropriate technologies to the private sector. Although technically part of the overall science and technology master business process, it did not associate the thrust of any effort toward support for its primary mission. It was, for all practical purposes, built as a stand-alone workload. Consequently it was given lower priority in a more highly centralized process. For business managers and technologists who traditionally look for ways to exchange information to help in their assigned work, this effort was seen as limited for their use and as a one way transfer of technology. As a practical planning and day-to-day business tool, and from an accountability standpoint, it was a secondary mission with only residual or collateral potential benefits coming back to the military.

Keys to success in cooperative ventures: the technology product must be useful, and the experience must benefit all the parties.

Organizations, in their efforts to make technology transfer and cooperative R&D work for them, should focus more in the beginning of their planning process on what they do best. This means concentrating on the first class technologies they need in order to support their own mission, and on technologies that complement their core capabilities and that are of real potential value for commercial application. It is good to remember that for successful cooperative ventures the technology product must be useful, and that the experience must be beneficial to both (all) parties.

The Air Force has recognized the value of a product orientation in its technology planning, and has begun forming integrated product teams within each principal mission area. Integration efforts are being extended to product development and weapon system management practices.

Product should be the "first filter" for the organization. It provides a less encumbered focus on both what is needed to serve the customer and what can be leveraged into a better or cheaper solution. A product focus, if nothing else, gets the right people engaged at the technology level. Additionally, the burden of sustaining a large and unattached staff to plan, work solutions, and make key decisions (such as funding) with little performance accountability and even less familiarity with the technology itself, can be avoided. A few companies and government organizations in our study have been very successful through their ability to focus their technology transfer and cooperative R&D program efforts on customer and product, as we have already seen happening in laboratories of the Department of Energy.

FEDERAL ACQUISITION REGULATION AND PROCUREMENT PRACTICE

Prior to the most recent acquisition reform act, the government was seen as having little ability to reduce technology cycle times. Part of the blame is put on bureaucratic requirements such as Military Specifications (MILSPEC) and Federal Acquisition Regulation (FAR), which were created with the intention of benchmarking performance, and ensuring fairness and non-preferential treatment towards all government contractors. The rationale for MILSPEC was to ensure performance, particularly in the areas of reliability and safety.

It is impossible to conduct a cooperative R&D or technology transfer activity without engaging the culture that has administered FAR and MILSPEC for more than forty years. Although cooperative R&D is not a procurement activity, it is often perceived in that light. The problems are not with the individual researcher, but with the system and the culture. Many industry researchers agonize over the slowness of the government acquisition process. Legal requirements and internal regulations have made quick decisions a high risk and thus improbable activity.

Recent acquisition reform legislation has streamlined research and development contracting. Whereas it once took years to move through the checks and balances of MILSPEC and the FAR, now the defense policy requires that a waiver be granted to any program manager who wants to use a MILSPEC in lieu of existing commercial performance specifications. This is a dramatic change from the past.

A waiver is now required if defense program managers wish to incorporate a MILSPEC in a procurement that can be filled by a commercially accepted specification. While this can streamline the procurement process, it can be difficult for government contracting officials who are risk averse by nature.

The complaints that are still directed at some federal laboratories include the assertion that not all have gotten on board the acquisition reform train. These organizations are still thinking and acting in a very "linear" fashion, wary about the uncertain destination. In other cases federal employees seem intent on going with the one-way flow dictated by government procurement processes. Common industry experience in cooperative R&D agreement negotiations includes instances where government lawyers warn of the "sweeping power" of the government, missing no opportunity to make explicitly clear "who is in control."

The bottom line, time to market, minimum cost, profit, and market demand are the operational catch phrases for industry. The increasing costs of research and development and threats from international competition have forced the government and industry sec-

In the private sector, the bottom line, time to market, minimum cost, profit, and market demand are the operational catch phrases.

tors to consider undertaking similar actions. As they strive to find a common cause to work together, they are struggling to overcome the effects of many years of institutionalized skepticism and distrust. The inability to see beyond the differences to the similarities has caused many technology transfer and cooperative experiences to fail.

Technology transfer policy and cooperative R&D initiatives have not broadly altered the old paradigm of industry versus government relations. As the interests of the two sectors converge more frequently, this will change. Cooperative R&D agreements are legally binding, but they usually only promise "best effort" because of uncertainty of resources on both sides. This gives rise to a number of issues of contention between would-be partners. The more important issues involve matters of intellectual property, indemnification, protection of proprietary material, and government licensing rights. These issues of contention have emerged from a rigid tendency to adhere to past ways of doing business.

As federal resources have diminished and technology cycle times have accelerated, government is recognizing the need for change. The serious changes to MILSPECs and the FAR, that were pending in 1995, have been adopted. This is a powerful indicator of the government's willingness to use commercial standards whenever appropriate.

Understanding each of these cultural considerations and imperatives is the first step toward building the bridges needed for successful partnerships in technology transfer and cooperative R&D. We will now examine more closely some of the more proactive steps that you can take in bridging the cultural gap and leveraging your resources. These include networking, use of bridging organizations, and training. Making initial contact with the right partner has been described often as the most difficult step in this endeavor.

chapter 4

Building Bridges

A growing body of evidence suggests that except in areas unique to defense, military technology is not much different from that developed in the commercial sector. Where the two diverge is in the specific application. The same basic science that has supported the nuclear weapons and stealth technology programs has applications for the commercial sector. For example, the sophisticated modeling and simulation techniques used in weapons design have shown applications for composite materials technology. Proponents of marrying the two technology bases argue that the needs in the defense segment are not inherently different from those in industry.

A partnership in cooperative R&D evolves in very much the same way as does a marriage.* In the

*A special acknowledgment is due to Lieutenant Colonel Jeanne Sutton, U.S. Air Force, who, as a student at the Industrial College of the Armed Forces, wrote an essay entitled "Marrying Commercial and Military Technologies: Strategy for Maintaining Technological Supremacy." The analogy between the CRDA relationship and a marriage used here was developed by LtCol Sutton. Her work appears in *Essays on Strategy X*, edited by Mary Sommerville, and published by NDU PRESS, April 1993.

courtship phase, the CRDA partners are searching for a compatible mate. Early in the relationship, similar to the first year of a marriage, the partners are sensitive to their mate's values and critical priorities as well as their own. Sooner or later, there will come arguments about money, a lack of communication, or how to deal with the partnership's offspring—in this case, the product of the CRDA in terms of intellectual property.

NETWORKING

Direct contact between the industry and government science and technology staff is the most effective way to initiate a cooperative venture.

The first step, and the one widely viewed as the most difficult, in starting a cooperative venture is the selection of a potential partner. Networking for industry and government is critical to this end. This chapter illustrates how to build the bridges between these two cultures that are so different in the way they perceive each other and in the way they operate. We begin by offering suggestions to simplify the networking process and making contacts between potential CRDA partners. A description of the various bridging organizations who broker such marriages is provided, followed by a discussion on the use of training as an effective bridging mechanism for those involved with brokering and participating in technology transfer and cooperative R&D partnerships. Finally, this chapter examines some of the innovative bridge-building steps that leading organizations have taken to position themselves for successfully conducting CRDAs.

For a technology manager interested in initiating a cooperative relationship, it helps to have a viable, broad network already established. The potential partner may simply call upon associates and friends for assistance in uncovering the location, expertise, and resources of a suitable counterpart. As we noted earlier, technology assessment is a critical component in determining the value of a potential cooperative R&D activity. Therefore, direct contact between scientific and technical staff is the fastest, most effective way to initiate a cooperative venture. The scientists are the people best able to determine technical merits and feasibility of a proposed project. Further, our research has demonstrated that cooperative ventures which are founded on personal relationships developed in the course of professional activities (engineering associations and societies, conferences and workshops, etc.) have a better chance of succeeding than those which are not.

Nonetheless, industry or government managers who lack an established network have available to them a host of options with which to find and assess potential contacts. There are many methods of establishing long-term scientific relationships which rely on personal initiative and varying degrees of unstructured interaction. Some which fall under the heading of technical interaction include personnel exchange, technical assistance, and troubleshooting with and for industry. Other popular and effective options for establishing contacts between industry and government are on-site visits and seminars. Moreover, many federal agencies and laboratories sponsor and participate in conferences which are advertised in trade magazines and journals published by organizations such as the newly consolidated American Defense Preparedness Association and National Security Industrial Associations (ADPA/NSIA).

Just as the defense industry is shrinking, so too are the professional associations that represent and support them.

Several organizations are singularly focused on creating bridges between industry and government. These groups, discussed in more detail below, offer access to structured, pre-formed networks. Other organizations, such as the Offices of Research and Technology Application (ORTAs) and technology transfer focal points, provide support to government laboratories and other product research activities in a variety of ways, in addition to their bridging responsibilities. The goal of these bridging organizations is to enable potential partners to find each other, conduct successful negotiations, and even provide the funding mechanism for a certain level of technology transfer program activity and worthwhile initiatives. Examples of independent, often nonprofit bridging offices that support both communities include the National Technology Transfer Center (NTTC) and Regional Technology Transfer Centers (RTTC) under NASA, the Federal Laboratory Consortium (FLC), and various state economic development agencies.

To invigorate technology transfer, technology managers in government must more effectively market their expertise, resources, and technology needs just as their counterparts in industry must communicate their technology needs. Bridging organizations further assist technology managers in advertising their offices' capabilities to the technology community. A network, whether previously established or produced by a bridging organization, is crucial to finding a suitable, cooperative partner.

USE OF BRIDGING ORGANIZATIONS

The Small Business Innovative Research and Small Business Technology Transfer Research programs, although similar, complement the SBA mentor program that matches small and large R&D companies.

Many industry participants, especially those from smaller companies, complain about the many levels of bureaucracy through which they must pass before they talk to someone who can help them. They often become disenchanted with cooperative ventures due to a single bad experience. By eliminating red tape, bridging organizations increase the chances that potential industry partners will successfully gain access to federal technology and resources. The Small Business Administration (SBA) sponsors a number of programs to assist smaller firms. For example, the SBA mentorship program matches small and large companies to help the smaller firm learn from and "capture" a share of business with the help of the larger, more established firm. The SBA strengthened its commitment to technology transfer with its Small Business Innovative Research (SBIR) and Small Business Technology Transfer Research (STTR) programs, both of which have a presence on the Internet.

Many bridging organizations publish guides describing their capabilities. Their resources often include mailing lists to update potential partners on activities and advances, access to the Internet and other databases to locate resources, expertise on technology areas, directories of federal laboratories, and points of contact. As would be expected in the present push towards the electronic "information superhighway," the availability, content, control, protection, and access to relevant technology information by those interested in exploring cooperative ventures is a fast moving train. It is marked by both hope that the task of matching technology needs with technology customers can be made easier, and by concerns about the desire to protect perceived competitive advantage represented by that information. In any case, becoming familiar with these bridging organizations can help in staying abreast of the latest capabilities and in using them.

Appendix F: Bridging Organizations summarizes the most popular and useful organizations, their function, resources, and on-line Internet addresses. Also, this book's Summary and Appendix H: Technology Transfer Related Internet Sites further assists the reader with guidance to better find and establish R&D partnerships. Many bridging organizations approach the technology transfer process differently, so it is useful to know where to go for what services.

By eliminating haphazard or blind contacts, bridging organizations increase the chances that potential industry partners will gain access to federal technology.

In the federal sector, a full-time Office of Research and Technology Application (ORTA) is required for all laboratories with 200 or more full-time scientists and engineers to provide technology transfer assistance. Because of the diversity of scopes, some of these offices have more expertise and resources than others. It is important that a potential participant understand and succinctly describe his or her own needs in order to optimally harness the capabilities of the ORTAs. Specifically defining one's technology needs is an important step in locating the bridging organizations which then find contacts who complement the capabilities and fulfill the needs identified. Many ORTAs and department level offices have 1-800 phone numbers and Internet Home Pages to facilitate communications. It is even possible to conduct on-line searches of their technology databases to look for partners. The Air Force's TECH CONNECT program and IR&D database are two examples of outreach services.

Many ORTAs and department level offices have 1-800 phone numbers and Internet Home Pages. Outreach services include the Air Force's TECH CONNECT program and IR&D database.

In summary, there is an art to using bridging organizations. The following steps are generally a part of techniques that have proven successful:

- Conduct a thorough assessment of technology needs.
- Identify specifically where or with what agency the required technology competency is likely to reside.
- Attempt to contact the specific laboratory or facility directly. This will substantially reduce the amount of time it will take to review your inquiry.
- Concisely draft what your interest and results are.
- Ask who within the laboratory or facility is an expert in the area. Attempt to contact that person directly.

- Establish a routine for follow-up.
- If your need is continuing, ask to be placed on a mailing list or e:mail.

We now turn to another important element of building bridges, namely the role and value of training to not only the initiation of the partnership, but also to its ultimate success.

TRAINING

Full attention by both communities to technology transfer training needs is time and money well spent if there is a good prospect for an improved product and a payback, in either profit or the leveraging of resources, through partnering and technology exchange. Training should be offered to a fairly high cross section of the organization; ideally, it should take several forms to obtain the maximum benefit and synergy from a corporate technology transfer and cooperative R&D strategy. This includes quality instruction that overcomes barriers built out of ignorance and promotes an essential understanding in those areas of cultural conflict addressed in Chapter Three.

For the government, technology transfer training must be made a priority to build and maintain an effective level of awareness and competence in both finding and performing as a good partner. Obtaining the greatest possible return on investment requires putting more emphasis on, and funding into, active recruitment of industry partners; it may well extend to efforts to make capabilities known to the commercial sector through good information exchange. Although the government cannot remain current on the technological activities of every U.S. company, it can keep industry better informed of available federal technologies, facilities, and expertise. Equally important, government must also initiate contacts on its own through an aggressive outreach program. Business development, commercial market awareness, trade association interface, and technology area joint working level planning expertise, all representing skills that have not been traditionally practiced within the federal establishment, must be acquired.

For industry partners, technology transfer might be argued to be merely an extension of normal day-to-day technology acquisitions, only with a different partner. However, it would be a mistake to take the attitude that no additional training is needed, that only a more finely attuned antenna to the federal technology opportunities is sufficient. It is true that if technology exchange with the government is a good thing, then the marketplace will make it happen. However, it is also true that learning more about a potential partner is not achieved without effort. More importantly, in the marketplace, success comes to those who learn quickly and thoroughly about their potential partners and how to overcome obstacles. In the critical area of communications, industry's lack of information undermines effective connections. Small businesses in particular often face this problem because they tend to be poorly informed about the available federal capabilities. They do not have the resources to hire personnel specializing in technology transfer. Training programs to get connected and to learn about working with the government can therefore be very advantageous, depending upon the prior experience of an organization, its competitive market, and its own organic capabilities and infrastructure.

Learning about a potential partner takes effort. Lack of information in the private sector undermines effective connections.

The federal agencies and laboratories can facilitate effective communication by maintaining efficient technology transfer offices. Technology transfer officers must be trained specifically to market technology transfer to industry. Ideally, new recruits for ORTAs should possess substantial industry experience and backgrounds in marketing. A typical frustration for an industry person calling a federal agency or laboratory is to be passed through several levels of the organization before he or she encounters an effective contact. This scenario demonstrates what ORTAs need to be aware of in the conduct of outreach activities to the commercial market.

One obvious message from this is that federal agencies must encourage their technology transfer offices to become more proactive; otherwise they may be neglecting potential sources of future cooperative partnerships and funding. The second message is that industry itself should learn more about the evolving federal and regional capabilities in this area and the ongoing initiatives for information exchange, several of which were discussed earlier in networking and bridging.

Government employees, particularly those who work within an ORTA, should be encouraged to attend university level courses and industry sponsored seminars in marketing. As the marriage of governmental and industrial practices evolves, there will be greater need for cross-training. Industrial employees would benefit from attending ADPA/NSIA Advanced Planning Briefings to Industry and the Spring FLC Conferences directed at outreach, and in some cases various short courses and seminars offered by institutions like the Defense Systems Management College (DSMC) on the topics of acquisition reform, understanding the Department of Defense 5000-series directives, and dual-use technology and defense conversion.

An increased emphasis in technology transfer in general and the use of CRDAs specifically uncovers the need for combined training of industry and government participants. Several organizations on the government side, and several others in the private domain, whose purpose serves the technology exchange community, could develop and provide this type of training.

The common theme is that training plays an important role in building the bridges necessary to satisfy overall business planning objectives through cooperative activities. It is an investment, but a worthwhile investment if a firm or agency wants to get beyond the heroic actions of the few isolated individuals and achieve the real corporate energy and capability of a fully empowered workforce.

For industry, we recommend a few, fairly simple, common sense steps to build and maintain bridges to potential partners in the public sector:

If your organization has developed a Web presence, you should get to know your Webmaster, not only for help in navigating the Internet, but also for putting information on it for others to use.

- Subscribe to government oriented technical journals
- Identify key government researchers using web sites, e-mail, and organizational directories being posted by nearly all government labs.
- Communicate with them periodically.

INTERNET SKILLS EVERY TECHNOLOGY TRANSFER PROFESSIONAL SHOULD DEVELOP

1. Learn to use a web browser. Know how to configure the browser and how to use the mail agent, newsreader, and bookmark and save features.
2. Master the use of at least two search engines. Learn to apply Boolean logic to reduce the number of irrelevant "hits" and focus your attention on the information that matters to you.
3. Complement your search engine skills with learning how to use a directory service or "switchboard." These tools will help you identify discussion groups, news groups, special interest web pages, bulletin postings, and addresses.
4. Build an Internet toolbox with utilities appropriate to your professional activities. Tools enable you to attach files to e-mail messages, encode and decode formatted files, connect to file transfer protocol (FTP) sites, and construct web documents.
5. Integrate your collection of Internet tools into your favorite word processor, contact manager, calendar, and project management software. Sort "bookmarks" by category, use them to find and/or verify information, and share them with colleagues and decision makers.

TECHNOLOGY LEVEL INTERFACE AND PEOPLE DETERMINE SUCCESS

Technology level interface facilitates opportunities to work closely and cooperatively with industry. The relationships enable government to gain contacts and networking experience, contributing to a closer understanding of technology developments and innovations, which then must be integrated and applied to government missions.

Marriages of potential partners in cooperative R&D seem, more often than not, to grow out of professional relationships at the technology working level, and almost by chance. In most government organizations we studied, there appears to be no truly effective systemic outreach program in place, although S&T managers are working hard to improve technology needs and capabilities information exchange systems and opportunities. We contacted several innovative leaders who had unique backgrounds in technology exchanges with industry and/or academia and who were adept in business planning and communications. The difficulty they faced in performing their program office duties was that their programs were austere, somewhat isolated from the technology product they served, and that they had little training in how to "clone" the set of skills required for program growth and success.

The following paragraphs provide some additional views on technology level interface and training collected by the *Technology Exchange* research team. A continuation of the compendium of best practices and lessons learned, begun back in Chapter One, will conclude this segment on bridge-building, and prepare you for a detailed look at the CRDA process in Chapter Five.

Community Views

A significant number of industry partners in our survey cited on-site visits and personnel exchanges as the optimal method for finding suitable contacts. They also stated that this form of contact created the most useful environment for future cooperative ventures. Some survey respondents, however, felt that personnel exchanges require too great an investment in time. Industry participants were especially skeptical about the potential benefits of exchanges outside the high tech or leading edge environment. The decision really comes down to weighing the costs of this type commitment against the particular needs of the individual organization.

The successful commercial firms and federal agencies in technology exchange that we talked to, and those who were posturing to take advantage of future opportunities, placed a heavy emphasis on the value of training in this area. Many of them have institutionalized their training program for production and staff activities to insure the right skills are involved in marketing, research, communications, planning, budgeting, technology interface, contracting, and legal. Some also contract for training or pursue formal training conducted elsewhere.

Making Contact, Selecting Partners, and Training People

Study participants were unanimous in discussing the importance of well qualified and trained people involved in the technology transfer and cooperative R&D program activities. Likewise, each mentioned the difficulty of getting started as perhaps the biggest challenge in the CRDA process. We found several examples worth citing as best practices in this area.

One excellent example of successful outreach initiatives comes from a leading Department of Defense logistics center. The Advanced Manufacturing Technology Center at McClellan Air Force Base, California, had integrated its technology transfer program efforts with its overall technology investment strategy to ensure the preservation of its "one-of-a-kind" capabilities and facilities for mission purposes in an extremely uncertain defense downsizing environment. The fear was that the draw-down, perhaps from base closure and program termination, would occur too fast to enable the organization to diversify its business process and maintain essential core capabilities. The Sacramento Air Logistics Center, through this active technology transfer office, was particularly effective in marrying its capabilities to the needs of state and local industry through the Federal Laboratory Consortium for Technology Transfer and other networking efforts. Cooperative partnerships evolved, and the result of the technology exchange was fulfillment of both military systems support and vital industry needs, identification of dual-use technologies for existing capabilities, and advances in technology.

Another aspect of Sandia National Laboratory's progressive program (discussed in Chapter One) demonstrates the importance of training to move technology transfer from theory to practice at the technology and product level. The scope and extent of the program at this particularly successful Department of Energy laboratory suggests that the integration of technology transfer into the structure of the organization was necessary for it to be effective. This laboratory developed a technology transfer training program for its employees, from senior scientists to students, on the requirements of technology transfer. Training like this can make technology transfer an integral part of the business and culture of the organization. In fact, in this case, it evolved into an "attitude" that technology transfer was part of everyone's job. This attitude and well trained people were considered prerequisite, and therefore responsible for the success of the organization during a volatile period of massive changes in its market.

A third example in this area, this one an outstanding business practice of a leading commercial industry firm and its Department of Defense partner, Rome Laboratories, Griffiss Air Force Base, New York, reiterates the value of technology information in accessible data bases, but in a way to protect as well as inform potential partners. The lessons learned and the reminder from this are that, first, caution is needed in the development of networking tools, and second, that time and training are needed to locate and nurture relationships. In highly competitive commercial markets, knowledge about the technology and the organization's interest in it is often very sensitive, even critical. Several of these industries are the type the government technology transfer programs need to attract. We saw in Chapter Three that the competitive edge is maintained through the quality of the technology product (i.e., intellectual property) and through timeliness in meeting market demands. This firm has been highly successful in protecting this competitive market edge, due in no small way to a sound industrial security program. It experienced further success in its ongoing cooperative initiatives, due to its ability to choose the right partner, in this case Rome Laboratories Photonics Center, that could meet its market criteria and demonstrate the necessary sensitivity to protect its competitive interests.

Other relevant observations and planning advice collected by the study team from organizational leaders on building bridges for successful technology exchange are summarized as follows:

- An effective government grassroots networking and outreach program was demonstrated on more than one occasion, and contributed greatly to ultimate success of technology exchange initiatives. Seeking a regional flavor and partnerships with both industry and academia brought a great deal of interest and diversity to the challenge. The consensus shared by all was that "getting started is toughest part."

- Industry experience showed that early contact at the technical level and some preliminary work with legal and contracts people were often the difference between success and lengthy delays in the negotiation process later. Those interviewed agreed that each government laboratory they dealt with was different, complicating the process for industry participants. The key ingredient is "attitude."
- From a program structure standpoint, the value of having the right people in the technology transfer program was emphasized. These people need to be experienced in the CRDA process and well grounded in both technical and business process. The focus of the partnership initiative needs to be on organizational objectives. Regular face-to-face contact with the laboratory and the technologist is also invaluable.

In summary, it would be fair to say that building bridges between two vastly different business cultures, whose shared interests compel them to work together,

Three elements of a successful program are: (1) networking to get started; (2) early team building efforts with technologists, legal, and contract support personnel; and (3) regular face-to-face contact throughout the project duration.

Industry	**Government**
• Visit facilities	• Employ those with business expertise
• Explore personnel transfers and exchanges	• Educate in marketing
• Become familiar with interface organizations	• Employ technology assessment
• Attend government conferences	• Market through trade journals technology specific publications
• Be patient and persistent in overcoming cross-cultural differences	• Send employees into the the commercial sector
• Build relationships with technical staff	• Provide for sabbaticals
	• Facilitate industry/industry association visits

Ways to build support bridges.

is the key first step for successful partnerships and in advancing the state of technology for the future. Like any trip, the placement of that first step determines the path a traveler will follow and the goal he will reach. Clearly, the best path is the one that promotes an enlightened joint approach to what has led to success in the past. Each must adapt. Fortunately, for the unseasoned or even the practiced voyager, the proficient use of networking methods, available bridging organizations, and training program resources can assist in making that first, important step.

Once a suitable partner with complementary needs and resources has been located, the two parties must then prepare to undertake the joint venture. This requires formulating a work plan and navigating the intricacies of legal negotiations. The next chapter offers specific advice on the CRDA process, from establishing a good work plan to concluding the negotiations and executing the CRDA. To supplement these discussions, this guide provides the advice of industry and government experts who have successfully completed several cooperative agreements.

chapter 5

The CRDA Process

This chapter goes into detail about the technology transfer process, using the CRDA mechanism as an example. Congress created the CRDA to simplify negotiations by making the vehicle exempt from the requirements of the Federal Acquisition Regulation (FAR). Essentially, the CRDA was designed to benefit industry by providing doorways to gain access to government facilities and scientists, and the technology developed there. This benefits the government by, ultimately, lowering the cost of items it needs from the production base. It is not meant to provide a vehicle for justifying funding to continue operating government laboratories.

As we guide the reader through the CRDA process from the initial work plan to final execution, we emphasize areas that pose typical obstacles and not-so-obvious surprises. The issues we raise in this chapter do not comprise an exhaustive list, but only those aspects we found to be most relevant. The reason the CRDA vehicle is singled out for in-depth analysis is simple—the CRDA is one of the most intricate of the technology transfer mechanisms to negotiate. Surprisingly, it is also popular. As it turns out, the CRDA is the most powerful and universal tool, accommodating hundreds of different versions and applications. A Model CRDA is provided in Appendix G.

The most important lesson the reader can learn from this chapter is that the work plan and final execution are just as important as preparation and negotiation. The energy required to complete the CRDA process, if it were graphed against time, would resemble a parabola. The start of the process is the work plan. It requires the smallest investment of energy. Following the work plan, and increasing in time and degree of energy required for success, lies initial preparation. The next major facet of the process is the actual negotiations, which also require significant effort to complete successfully.

The CRDA was designed to provide doorways for industry to access government facilities and scientists, and the technology developed there.

The last stage in the CRDA process is execution, which requires the least amount of effort or energy, but the greatest resource input. Nonetheless, execution is arguably the most important facet of the process. Like a good golf swing, proper execution of a CRDA extends all the way to follow-through. Research has shown that neglect at this final stage can scuttle the project as thoroughly as a sloppy work plan can in the initial stage.

The final section of this chapter highlights specific lessons learned from our research. It is presented as a series of vignettes, each with a moral for others to glean. This section will benchmark both good and bad experiences to highlight success and failure at the various stages.

THE WORK PLAN

Negotiating agreements between the public and private sectors is a complex process. Both government and industry participants must identify their objectives and assess the resources and requirements of the other party. The work plan serves as a crucial, initial step toward effective negotiation. This plan describes the aspects of the operating environment in which public and private organizations must interact. It also contains a detailed exploration of the means and ends of a cooperative venture.

A work plan that is specific and detailed can help to greatly simplify the later phase of negotiations.

The better and more detailed the proposed work plan, the more effective the negotiations will be. The construction of a detailed work plan forces venture partners to evaluate those aspects of an agreement that are most important to each organization and those that are superfluous and irrelevant. For example, if the intent of the CRDA is to conduct basic R&D, then concerns over product liability will be less important than, say, ownership of the research data. On the other hand, if the intent of the CRDA is to market a specific product, then issues of third party licensing, publication of research results, and royalties from production of any new technology will be significantly more important.

A second benefit of developing a detailed research plan is that it helps with the "inside sales job." The CRDA must be sold to the participants' own organizations. Garnering internal support in areas of potential uncertainty and technical risk is difficult. A simply stated objective and a detailed work plan assist all participants in gaining organizational commitment.

A work plan enables parties to agree to milestones and accountabilities before the actual CRDA instrument is drawn up.

Survey respondents and, privately, many experts suggest that it is counterproductive to bring legal representatives into this stage of the process. There is a tendency to muddy the water by haggling over details that may be unimportant from the start.

Issues raised in the work plan must deal with factors that are advantageous to all parties in order to ensure continued cooperation. Discussions within the government laboratory must focus on the relevance of the proposed CRDA to the laboratory mission, the benefits to and needs of the laboratory, and the federal resources required. The industry participant must discuss the company size and resources; potential profits; its ability to develop, manufacture, and distribute new products; and the need to protect proprietary data and obtain a competitive advantage. Finally, both parties should be concerned with the technology's stage of development, the resources required to commercialize the technology, additional required expertise, potential fields of use, and the size of the new technology's market.

The work plan should show what kind of homework you have done. It should address all factors that make the arrangement a win-win situation.

While having a detailed research plan in hand in no way ensures that the CRDA will be successful, the greater the understanding of the goals and visions of the participants for the contemplated joint venture, the more likely it is that the negotiation process will stay on track until success is achieved. Those who approach the government laboratories with ill-defined requirements or needs will find the negotiations unduly difficult and cumbersome. The lesson learned here is if you want streamline the negotiation process, you need to do the homework first.

PREPARING THE CRDA

Once the work plan has been set down in detail, the next step toward a cooperative R&D agreement is to prepare the CRDA. Even at this prenegotiating stage, two distinct perspectives come into play.

Government Perspective

Prospective CRDA participants from industry should be aware of specific concerns of the federal laboratories that potentially hinder negotiations. The following is a detailed list of the most common government issues that industry must tackle during negotiations.

Fairness of Opportunity

Federal law requires that opportunities to work with government laboratories be as widely advertised as possible. Furthermore, when a government agency agrees to grant exclusive licenses on inventions made outside a CRDA, it is required to advertise that intent prior to granting the license. It is important that government laboratories avoid even the appearance of showing favoritism to an individual or particular organization.

The Federal Technology Transfer Act (FTTA) of 1986 (PL99-502) states that preference should be given to businesses located in the United States, particularly companies that agree to manufacture the technology substantially in the United States. To ensure maximum benefit for the taxpayers' investments in R&D, government agencies are required to certify that U.S. business and U.S. workers receive preference in the commercialization of the technology.

Special Consideration for Small Businesses

For most categories of interest, a small business has 500 or fewer employees. The FTTA mandates that special consideration be given to small businesses because they employ the majority of U.S. workers and are often more ready to accept risk and to innovate than are larger companies. Moreover, small companies do not generally have the R&D funds and other capital resources required to effectively commercialize technology.

Conflicts of Interest

Prior to and during the negotiation process, federal laboratories must be familiar with local conflict of interest policies. As stated above, these laboratories must avoid even the appearance of impropriety in their dealings with private sector parties. Generally, conflict of interest provisions regulate the employment of federal workers by private sector participants and the receipt of gratuities. These provisions protect against situations where inventors receive "undue gain" as a result of the positions they hold in the federal government.

Freedom of Information Requests

The Freedom of Information Act (FOIA) of 1986 (PL89-554) allows public access to public records including some results of federally funded R&D. When considering cooperative research and development, private sector participants are unlikely to invest in the commercialization of a technology if their competitors would have full access to their research. Certain provisions have been made in the cooperative R&D legislation to protect trade secrets or commercial or financial information developed under a CRDA from disclosure for up to five years (see, for example, the discussion below regarding trade secrets). This applies to data not otherwise protected by patent, copyright, or some other vehicle. The Technology Transfer and Advancement Act of 1995 directs the government to not publicly disclose trade secrets or financial information that is privileged or confidential; thus it is important for private sector participants to clearly label their proprietary material. For most instances where proprietary material is shared with federal laboratories, the simple marking of the material as proprietary invokes the special rules for handling. The marking typically runs across the bottom of the page, and the instructions (e.g., "Information on pages marked 'Proprietary Data' may not be disclosed without prior written approval.") are provided as part of the front matter. For most instances where proprietary material changes hands, the simple marking of the

Proprietary materials shared under a CRDA may be marked as privileged and confidential material to protect against disclosure.

material as proprietary invokes the special handling rules of the CRDA.

Industry Perspective

Like government, industry has some specific concerns that need to be recognized within the work plan and during the negotiations. These are described in the following paragraphs.

Risk versus Return on Investment

When negotiating an agreement for a new product or service, private sector participants must weigh the potential risks of the situation. A low risk opportunity involves a product that enjoys widespread demand, a low R&D investment to bring the technology to market, and a high potential for return on investment. Conversely, a high risk situation might include the development of a new technology, the requirement of a substantial investment to market the technology, and a low potential return. In order to weigh potential risk versus potential return on investment, private sector participants must ascertain the pace of technology (when will the product become obsolete), its cost of development, the company's ability to quickly increase product sales, its ability to obtain capital, and the cost of that capital. Finally, these factors will all be contingent on commitment of the federal parties as negotiated in the CRDA.

Negotiation Response Time

Now that the novelty has worn off, many government agencies process a CRDA in a number of days versus number of weeks.

The private sector often views the government laboratories as slow and bureaucratic. This is not true of all government laboratories, but the slowness of the pace is especially notable when compared to the rapid speed industry strives for with most technology developments. Industry participants are concerned with introducing new technologies into the marketplace in the shortest time possible. Clearly, the duration of the negotiations is a factor in this time horizon.

Some laboratories appear to be reaching saturation in their ability to respond to CRDA requests. One federal laboratory, for example, is only able to accept one out of four CRDA requests because of lack of adequate funding for technology transfer activities. This is a particular frustration for companies that expend resources to prepare for CRDAs, only to find that the projects must be postponed for lack of government funding.

Finally, non-uniform governmental policies and procedures often place a burden on industry. These areas include indemnification, ownership of intellectual property, and accounting standards addressing the scope of government audit privileges. The various Departments (Energy, Defense, Agriculture, etc.) and their operations (Government Owned, Government Operated versus Government Owned, Contractor Operated) are so different that it is hard to mention them in the same discussion. This lack of uniformity contributes to the delay and frustration.

NEGOTIATIONS

This guide cannot attempt to offer a meticulous, step-by-step procedure for negotiating CRDAs. Instead, it discusses in detail potential surprises and obstacles in successful negotiations. There are five principal areas of concern at this stage: (1) product liability and indemnification, (2) intellectual property rights, (3) licensing, (4) publication rights, and (5) dispute settlement and CRDA termination. This section not only describes the various potential hurdles in these areas, but also suggests proven methods to circumvent them.

As mentioned above, the FTTA mandates that preference be given to companies that "substantially" manufacture CRDA inventions in the United States. First, it should be noted that this is only a "preference." The government can go to a foreign manufacturer if there is no U.S. producer. Second, the term "substantially" is not clarified within the law. Industry compliance with this rule could result in competitive disadvantage in global manufacturing as production shifts away from low cost sites. Multinational corporations would prefer a loose interpretation of this requirement, which currently prevents many such firms from participating in CRDAs. Furthermore, this law may force industry into non-compliance with foreign content laws.

To avoid this problem, the Department of Defense interprets "substantially" as *by rule of reason* or as *with regard to circumstances*. The Department requests that CRDA participants demonstrate that they will manufacture resulting products as much as possible in the United States. Industry partners must justify their position with data from manufacturing operations. It is important to note that no CRDAs initiated by the Department of Defense have stalled due to U.S. manufacturing content laws. Moreover, only a few CRDAs have been signed that do not actually increase U.S. manufacturing. In these instances, the CRDAs were approved in the belief that domestic technology gains would more than offset any loss to overseas manufacturing.

Product Liability and Indemnification

Based on a series of interviews, we discovered that government insistence on indemnification often created obstacles to successful CRDA negotiations. This indemnification covers situations spanning from use of research tools during R&D to potential law suits that may result from the sale and use of the developed products.

The federal government accepts royalties for use of patents but, nevertheless, insists on being indemnified in future law suits from damages that may result from the sale and use of products. The government argues that this position is fair because it has no control over engineering and design decisions made by companies after the CRDA is completed. Companies argue that this decision suggests that the government wants benefits without taking its share of the risk.

The National Institute of Standards and Technology and the Department of Defense sometimes grant their company partners indemnification from law suits involving third parties licensed by the government. This lets companies acknowledge that the federal government has little control over development work conducted subsequent to the termination of the CRDA, and lets the federal government acknowledge that companies have little control over government decisions to grant licenses to third parties. In the future, the U.S. government may insist that all third party licenses indemnify both the government and the original CRDA partner(s). Also, industry may be able to limit its liability to some fixed percentage of the value of the CRDA.

Intellectual Property Rights

Intellectual property is a generic term that applies to any invention, discovery, technology, creation, development, or other form of expression of an idea that arises from research activities, whether or not the subject matter is protectable under the patent, trademark, or copyright laws. The right to intellectual property can be bought, sold, leased, rented, or otherwise transferred through contracts or licenses. If intellectual property rights are not adequately considered throughout the cooperative R&D process, valuable opportunities may be lost and serious liability issues may result. Intellectual property rights are protected through the use of the following devices: patents, copyrights, trade secrets, and trademarks.

Rights to intellectual property can be sold, leased, rented, or otherwise transferred through contracts and licenses. These rights are fully negotiable in the CRDA process.

The transfer of intellectual property rights affects the marketability of a product and the selection of the appropriate manufacturer. Therefore, the right to intellectual property is often a substantial component of the cooperative research and development process. Hence, it is of paramount importance during CRDA negotiations.

Invention

To qualify as patentable subject matter under a CRDA, it must first qualify as an invention made under the CRDA. When used with the invention the term "made" [15 USC 5703(10)] means "the conception or first actual reduction to practice of such invention." Boilerplate CRDA language typically provides the government with at least a nonexclusive, irrevocable, paid-license to practice the invention or have the invention practiced throughout the world by or on behalf of the Government.

For something to qualify as an invention under a CRDA, it must be conceived or actually demonstrated as workable by the inventor (this is the reduction to practice referred to above). If the invention is conceived prior to the CRDA, but reduced to practice under the CRDA, the government may obtain a royalty-free license to use the invention for government purposes. CRDAs can be entered into with inventions that are not reduced to practice prior to the CRDA and still be beneficial to the companies entering into the CRDA. This should be factored into the evaluation process when companies determine whether they want to ensure that prior inventions are reduced to practice before entering into a CRDA.

In addition, the government may claim improvements to current or parallel inventions and assert rights to these inventions. The industry participant may want to specify an exclusive license to any joint or government patent.

Patents

A patent, for our purposes, is a contract between the U.S. Government as represented by the U.S. Patent and Trademark Office and an inventor whereby the government gives the inventor the right to exclude others from making, using, or selling the invention for a period of 17 years in exchange for the inventor's complete disclosure of the invention (effective June 8, 1995, under most circumstances, the length of a patent will become 20 years from the date of filing the application). CRDAs typically give the party with ownership interests in any invention six months to file a patent application before the other CRDA partner may file. Patentable subject matter falls in four classes:

1. *Process:* a method, a series of actions or operations achieving a physical or chemical change in character or condition of an object. A process may be chemical, mechanical, electrical, or other. A process may also be a new use for an already existing product.
2. *Machine:* any apparatus having an assemblage of parts that function in conjunction with one another.
3. *Manufacture:* anything made from raw materials.
4. *Composition of matter:* anything resulting from chemical compounds or mixtures of substances that have properties different from those of its individual ingredients.

Federal government agencies and departments file for and maintain patents for the following reasons:

- Patents provide the primary legal protection for the fruits of Government-funded research.
- Patents afford the best form of defense against others patenting inventions which the Government has already funded.
- Patents provide the highest form of recognition in the scientific community for the accomplishments of technical personnel.
- Patents provide a source of income to the government by the licensing of patent rights to the private sector.
- Patents advance the progress of the sciences by establishing a vast portfolio of the latest advances in technology.

In the case of government owned, contractor operated (GOCO) laboratories, the operator of the laboratory may obtain commercialization rights to these inventions under the prime contract or by waiver of the government's ownership rights. This is cause for some concern to other private sector companies who wish to engage in cooperative R&D with one of the GOCO laboratories. If the CRDA is not properly written, questions may arise as to whether the private sector company who operates the laboratory owns the proprietary rights to technology developed at its facility even if the operator did not bear any of the R&D burden in developing the new technology.

The Technology Transfer and Advancement Act of 1995 states any royalties or other payments received from the licensing and assignment of inventions shall be retained and disposed of by the laboratory which produced the invention as follows. Each year, the inventor or coinventors is entitled to the first $2,000 and thereafter at least 15 percent of the royalties or other payments transferred to the laboratory. Furthermore, appropriate incentives may be provided to laboratory employees who are not inventors but who substantially increased the technical value of the invention. Royalties received per year by any individual cannot exceed the $150,000 limit without Presidential approval.

Inventors are entitled to the first $2000 and thereafter at least 15 percent of royalties received each year from any patent licensing.

The discussion of patents above should make it clear that allocation of ownership and license rights are fully negotiable. The greater the percentage of costs funded by the industry participant, the stronger the claim may be for a larger share of rights in patents produced under the CRDA.

Industry participants also may be involved in CRDAs which produce patentable technology principally suitable to government uses. Despite the funds the commercial company had invested in a cooperative venture, it cannot charge royalties to the government for the use of the subject technology in a different contract because the government retains license. Competitors who win government contracts will be entitled to use the technology under the government paid-up license without payment of royalties to the initial CRDA industry participant. However, granting to government a royalty-free license does not deprive the industry participant of any competitive benefit when bidding for a subsequent government contract. Furthermore, the industry participant can offer resultant technology to the government in a different contract, even if the government retains a royalty-free license.

Disclosure of Patentable Technologies

An invention that is determined to be patentable must be new, useful, and non-obvious. Scientific laws, physical phenomena, or things occurring naturally in nature (such as plants in the wild) cannot be patented; however, under certain conditions, patenting computer software is possible. When modifying a proprietary software application for the government, the developer would protect the base code, and license the modified code for restricted use so that it is not reverse engineered for other purposes. Premature public disclosure of a new technology may undermine the benefits which emanate from a patent. Patent review should be obtained from patent counsel to protect the invention before it is publicly disclosed.

Statutory Invention Registration

The Statutory Invention Registration (SIR), created by the Patent Law Amendments Act of 1984 (PL98-622), provides prior art to the Patent and Trademark

Office which may be considered in evaluating the patentability of inventions. The requirements to apply for an SIR are similar to those for a patent application. However, unlike with a patent, the applicants receive no protection on their inventions. The SIR does not allow the inventor to enforce infringement or collect damages. Others may make, use, or sell the invention without needing a license.

Copyrights

A copyright is defined as an exclusive right granted by the U.S. government to the authors, composers, artists, or their assignees. The life of a copyright is for the life of the author plus 50 years; for a work-for-hire and anonymous works, the life is 75 years from publication or 100 years from creation, whichever expires earlier. Copyrights are always handled separately from patents.

Copyright protection is initiated with the creation of a work. Registration of copyrights with the federal government is optional and is only required in order to receive statutory damages. A work can be registered by submitting an application, one copy of an unpublished work or two copies of a published work, and a check to the Copyright Office within the Library of Congress.

Copyright protection is not permitted for any works prepared as part of the duties of government employees. However, the government may receive and hold copyrights transferred to it by assignment or bequest. The copyrighted works of GOCO employees are generally regarded as works for hire and the copyrights are, therefore, owned by the commercial operator. Government approval is usually required to enable the GOCO to enforce the copyright and to affect licensing arrangements and commercialization.

Trade Secrets

A broad definition of trade secrets encompasses virtually any compilation of business information that provides a competitive advantage. More narrowly defined, "trade secrets encompass commercially valuable plans, formulas, processes, or devices that are used for the making, preparing, compounding, or processing of trade commodities and that can be said to be the end product of either innovation or substantial effort."* The FTTA directs government laboratories to refrain from disclosing trade secrets of all collaborating parties obtained in the conduct of research or as a result of activities under a CRDA. This exclusion makes trade secrets extremely important to the competitive position of most companies engaging in cooperative R&D.

Trade secrets can be an alternative way of protecting subject matter which can be either patented or copyrighted. It is used frequently for information that would be compromised if made publicly accessible. However, trade secrets may be effective protection only if the information is not subject to reverse engineering.**

Unlike patents and copyrights, which are protected by federal laws, trade secrets are protected by state laws. They can be defended in court if they are actually secret, substantial, and valuable. However, it is important to note that because data generated at federal facilities are subject to public dissemination, the data may not qualify as trade secrets.

In November of 1989 the Federal Technology Transfer Act (FTTA) was amended in order to prevent the disclosure of trade secrets or commercial or financial information that is privileged or confidential under the meaning of Section 552(b)(4) of Title 5, United States Code, obtained from a non-Federal party while conducting research under this Act and participating in a cooperative research and development agreement. Also, the Government will protect against dissemination, for up to 5 years, information developed as a result of research and development activities conducted under this Act if that information would be a trade secret or commercial or financial information that is considered privileged or confidential if it had been obtained from a non-Federal party participating in a cooperative research and development agreement. After the fifth year, the industry party must assert that the information is under exemption (4) of the FOIA. This exemption protects commercial and financial information and trade secrets obtained by the govern-

---

*This formula for defining trade secrets was adjudicated in *Public Citizen Health Research Group v. FDA,* 704 F. 2d 1280, 1288 (D.C. Cir. 1983).

**Reverse engineering is dissection of a product to see how it was constructed.

ment during the CRDA from public disclosure if there is a *compelling* reason for the agency or contractor to maintain secrecy. Even after the five year moratorium on FOIA disclosures has expired, the government is obligated to inform the industry participant of all FOIA requests *and* the government's intent to comply. The industry partner to the trade secret may then wish to take additional steps including injunctive relief.

Confidentiality Agreements

When negotiating the CRDA, the industry participant may include a confidentiality agreement as a further protection against improper disclosure of proprietary information including trade secrets. This agreement must specify data considered protected by both parties, set explicit limits on the use of such data, and identify remedies in the event of a breach of the agreement. As this area is already protected by law (see discussion of Trade Secrets above), the separate agreement constitutes extra work for the legal department.

Within the confidentiality agreement, some remedies for improper disclosure might include a contractual acknowledgment of the right of each party to seek injunctive relief. However, since the most important form of protection is the patent, companies generally seek to obtain title to intellectual property (IP) for commercial purposes.

Ownership of IP provides significant incentives for the commercialization of government technology since commercialization is unlikely to occur without title. (The reason is quite simple. Industry will need to recoup R&D costs through licensing of the technology and royalty payments if not through actual production and marketing of the new technology.) Further, ownership of IP prevents the commercial use of the technology by a company's competitors since the government cannot license that technology to a competitor.

Trademarks, Trade Names, and Service Marks

The terms trademark, trade name, and service mark are used similarly. The purpose of this protection is to establish a word, name, symbol, device, numeral, picture, or any combination thereof, in any form or arrangement, for use by a party to uniquely identify the origin of goods or services.

A trademark is established by actual and continuous association with products or goods in interstate commerce. Trademarks may be registered and protected for exclusive use. Even if a trademark is not registered, it is still protected to some extent. Ordinarily, there is no reason for the government to have any right to trademarks related to inventions. The industry participant to the CRDA should generally negotiate sole ownership of all trademarks, trade names, or service marks.

Standard Proprietary Rights

Three standard clauses cover the handling of proprietary rights. The first two, unlimited rights and limited rights, pertain to procurement contracts and, consequently, are not discussed in this guide. The third type, government purpose license rights, are encountered in cooperative R&D agreements.

For the purposes of a CRDA, government purpose license or GPL means a license to the Government conveying a nonexclusive, irrevocable, worldwide, royalty-free license to practice and have practiced an invention for or on behalf of the Government for government purposes and on behalf of any foreign government or international organization pursuant to any existing or future treaty or agreement with the United States, and conveying the right to use, duplicate, or disclose copyrighted works or proprietary information in whole or in part and in any manner, and to have or permit others to do so, for Government purposes. Government purposes include competitive procurement, but do not include the right to have or permit others to practice an invention or use, duplicate, or disclose copyrighted works or proprietary information for commercial purposes.

Protection of Proprietary Information

Proprietary information and data from the industry participant may be provided to the laboratories and other facilities in the process of cooperative R&D. Such information and data require protection in order to retain their commercial value.

To be protected as proprietary information, the participant must have developed the information in question at private expense and must clearly mark that information as proprietary. A notation like "Proprietary Data" is typically placed at the bottom of the page. Under the rules of the FTTA, proprietary information that qualifies and is well marked as such must be maintained as confidential indefinitely.

The period of confidentiality will terminate when it can be demonstrated that proprietary information is no longer confidential. Furthermore, all participants must ensure that the CRDA contains language that arranges for the disposal of proprietary information, addressing whether the information should be returned, destroyed, or retained for future use.

In the case of CRDAs with GOCO laboratories, government agencies are not directly party to the CRDA. The industry participant should negotiate for an express confidentiality agreement with the agency or, in the event that the agency refuses, the CRDA should bind the government to the language already set forth in the standard agreement.* Typically, government relies upon the Trade Secrets Act which subjects government employees who improperly disclose proprietary information to criminal sanction. [See the Trade Secrets Act (18 USC § 1905)].

The party to a CRDA who generates protected information is normally given the absolute right to designate that information as proprietary and therefore protected. However, either party may request that information developed by the other party be designated as protected.

Negotiations must generate a policy that protects proprietary information and establishes guiding principles for carrying out that policy. There are several guiding principles with this intention:

- Only those materials or data that are absolutely essential to fulfilling the project's objectives should be surrendered.

*Most boilerplate agreements possess language concerning confidentiality. Should the agency refuse to incorporate a separate confidentiality agreement, the industry participant should insist that the government agency is listed as a party to the CRDA in the case of CRDAs with GOCO laboratories.

- Only those individuals who have a need to know and are engaged in essential activities for the project should be informed of the nature of the material.
- Careful consideration should be given to where the proprietary information is stored and how it is subsequently accessed.
- Under no circumstances should data be orally transmitted unless it is promptly reduced to writing by the owner or sponsor and appropriately marked with a legend that specifically identifies the restrictions for use and disclosure of the information or data.

Rights to Application Beyond the Intended Fields of Use

Smart bomb technology has extended into a medical application in the location and treatment of cancerous tumors.

Products developed under a CRDA may have other applications. The question of whether industry (and government in some cases) can utilize the new technology outside of the original intent of the original CRDA research plan and the work plan must be considered during the negotiation phase. One aspect of "smart bomb" technology has already found application beyond the intended field of use. Algorithms developed as part of automated target recognition (ATR) capability of "smart bombs" now help doctors locate and treat cancerous tumors.

Licensing

Licensing is the transfer of less-than-ownership rights to another party so that it may use the intellectual property. Usually, patents are secured to prohibit competitors from using an invention. Licenses are negotiated when the invention is to be applied to a commercial product by a party that does not own the patent.

Licenses transfer less than ownership rights, and carry their own restrictions, such as exclusive and non-exclusive use.

The granting of a license may be exclusive or non-exclusive, restricted to a particular field of use, and/or restricted to a particular geographic territory. Private sector participants generally consider an exclusive license preferable because it restricts the competition from utilizing the technology developed. These participants are sometimes amenable to acquiring a license that is restricted to a particular field (e.g., the original intent of the technology as developed by the cooperative R&D) or geographic territory in which their company operates.

Non-exclusive licenses tend to be less agreeable to the private sector, because they allow other companies to acquire licenses to the same technology *and* market a similar product in the same geographic area or restrict its field of use. Clearly, licensing is of paramount interest to both parties as government may tend to emphasize unrestricted licensing to promote royalties or build a strong indigenous industry, while private companies require a secure niche to effectively capitalize on their R&D investment.

Licensing from the Government to the Private Sector

The government may license its own inventions. Before it grants a license, it may require that potential licensees present plans for commercializing the invention. When granting exclusive licenses, if the invention to be licensed was not made under a CRDA, the government must publish notice of availability and notice of intent to grant an exclusive license and provide an opportunity for the public to respond. Geographic exclusivity allows more than one exclusive license to be granted per patent as long as the licenses are for different regions. Whatever arrangement is made, under a CRDA the government must retain a non-exclusive, royalty free, paid-up license for government use of the inventions.

Licensing from the Private Sector to the Laboratory

The government may acquire licenses to software and other intellectual property through CRDAs that specify limitations concerning use, copying, transfer, and disclosure. All laboratory employees are bound to follow these agreements and should ensure that information received is clearly marked with any restrictions that apply. The laboratory and the individual may be held liable for violating the terms of the intellectual property agreements. The government routinely conducts audits to ensure that unauthorized software is not being used on government equipment. The government can also obtain assignment rights from its contractors on contractor-developed technology from federally funded projects.

Publication Rights

This is a key issue that concerns industry. If research is prematurely published, industry can lose benefits accrued through its investment in the CRDA. Industry has a definite interest in ensuring that information does not leak to the public domain before a product makes it to market successfully. On the other hand, government often justifies its budget by publishing reports of its activities. The negotiation process must resolve this conflict.

Dispute Settlement and CRDA Termination

Negotiations should also develop a dispute settlement regime. Specific recourse to any disputes arising from the maintenance of the CRDA must be established so that minor problems do not develop into major stumbling blocks. This regime must address whether research is continued while the dispute is resolved.

The final step of the CRDA, termination, must be discussed during the negotiations. This may include specific objectives and assumed time horizons. This discussion should also address the conditions of termination should either party become dissatisfied with the operation of the CRDA.

EXECUTING THE WORK PLAN

J. David Roessner[2] claims that ". . . companies tend to interact with federal laboratories for reasons that have more to do with long-term, less tangible payoffs than with expectations of short-term business opportunities or technology commercialization." Our research generally confirms this conclusion; however, as we spoke to more and more people associated with firms which had successfully negotiated and executed CRDAs, we found that the level of sophistication as well as the expectations of all parties increases over time.

From companies that had experience working with public sources of technologies come the following insights and recommendations:[3]

- Establish a set of research objectives, timetables, milestones, and measures of success that is consistent and based on realistic expectations.
- Carefully select the public source with an eye toward relying on more than one or two sources for knowledge.
- Develop personal relationships between bench-level scientists and engineers, contract administrators, and project managers.
- Communicate through frequent project reviews tying continued investments to milestones and objectives.
- Be patient—projects may take longer than originally planned.
- Establish realistic exit criteria (to handle R&D failure).
- Maintain continuity (applicable to both sides).

BEST PRACTICES AND LESSONS LEARNED

Our compendium of best practices and lessons learned, begun in Chapter One, is continued here to acknowledge the importance and synergistic impact of such key elements as integrated strategy, program structure, goals, networking, and training (discussed earlier), as well as incentives, metrics, and leadership (addressed in the next chapter) on the ultimate success of the CRDA process itself. When it comes right down to making *Technology Exchange* work, the CRDA dynamic is really where the "rubber meets the road." The following first-hand vignettes on the CRDA process, shared by selected key leaders in successful enterprises, both public and private, continue to build our knowledge of what works and what does not work in technology transfer and cooperative R&D. Common themes, reflecting the general state of the community studied, are included along the way for perspective and comparison.

Interface at the technical level early in the process has been cited as a key element of success.

Effective Work Planning & Clear Objectives

The time to bring in the administrative support for help in preparing CRDA negotiations is after the technologists have a firm grasp of each other's needs and capabilities.

We have emphasized that the success of any technology partnership depends upon a clear view of organizational objectives and a good understanding of the resources and needs of the other party at the beginning of the CRDA process. The importance of technical level interface early in the relationship proved to be the key to the overall success of the CRDA experience for one defense industry firm in particular. In this case the technologists had built a thorough appreciation for each other's needs and capabilities before support offices (i.e., the lawyers) were involved and CRDA negotiations started. This provided a solid basis to proceed, fostered an element of trust between parties, and secured an attitude of joint commitment to making the process work.

A second defense industry firm benefited from the same technology level approach, but also found that preliminary work with government legal and contracting staffs was equally helpful in avoiding conflict later in the process. This was especially relevant to this firm since it had initiated partnership ventures with several government R&D organizations and had encountered varying, sometimes agonizingly cumbersome, procedures during negotiations with each.

One more illustration of the successful CRDA planning approach of the defense firms cited above was their active, albeit cautious, early involvement with the relatively new enabling technology transfer policy. These organizations correctly judged the value of continuity on both sides by maintaining contact with the government early in the process through CRDAs, given the rapidly changing defense industry environment and the severe impact it was having on new procurement contracts. Close working level relationships and an on-site presence with potential partners "would be meaningful downstream," they reasoned, ostensibly at the competition and production phase. The prevalent theme among participating defense companies was to maintain core competencies in potential lucrative technologies and to stay in touch for procurement contract potential with the ultimate customer.

A further significant aspect of effective work planning by each of these companies was in limiting their near-term objectives to lower the risk of cooperative R&D ventures. The first firm did so very effectively through participation in a large consortium with a government (Department of Energy) laboratory and 10 to 12 other industry organizations, both engineering and manufacturing. It leveraged shared resources to accomplish competitively safe, but labor-intensive, up-front research in a technology growth area. The other firm was able to advance or fine tune its technology research and focus through several cooperative technology exchange relationships. It found the CRDA approach easier to justify to internal management than "another marketing trip" and in showing tangible results from staying abreast of defense customer methods and state-of-the-art technology needs.

The Community Today: We found from industry participants a preference to use in-house R&D where possible for strategic ventures, and to employ external resources for lower risk, enabling technology (i.e., test capability, facility and equipment use, etc.) initiatives. Those studied were not "betting the farm" on technology transfer or dual-use activities, but were posturing for parallel, higher risk work envisioned in house. Defense industries in particular were staying inserted to see where government was going in order to compete for future procurement contracts. From government participants we found that bringing outside (i.e., industry or academic) energy into a cooperative research effort with shared facility, equipment, or process was welcome. They viewed it as a way to ultimately enhance mission supportability, both directly and indirectly, and in the long run if not necessarily in the near term.

Industry participants tend to use in-house R&D for strategic ventures and external R&D sources for initiatives that involve lower risk or provide access to enabling technology.

Model CRDA Tools & Practices

Much of the research on preparing the CRDAs pointed towards the adequacy or availability of model CRDAs for use by all parties. The improved modular Department of Energy CRDA instrument and the model CRDA developed by the Air Force (see Appendix G) for use as a template are good examples of the efforts being made to facilitate the negotiations. The following vignettes illustrate the value of a simple, flexible, and perhaps jointly developed template that can be tailored to meet specific needs.

An imaginative cooperative initiative to overcome legal impediments with the CRDA process was the use of a legal teaming arrangement to conduct educational seminars on cooperative R&D and technology transfer. The example we cite paired an expert from the Electronics Systems Center (ESC), Hanscom Air Force Base, Massachusetts, with a noted industry consultant. This forum was effective in educating would be participants from both sides on the legal and contractual nuances of the CRDA process. Perhaps equally significant, it heightened the overall awareness to the differences from the historical FAR-driven acquisition process, and promoted a spirit of joint cooperation.

One step suggested by a defense industry cooperative R&D manager was to encourage government to seek joint participation early from industry representatives in the preparation of model CRDAs. Another innovative firm in commercial industry observed that the best model CRDA was a simple basic framework agreement that addressed only the essential topics applicable to any cooperative partnership. The topics of special concern or applicability to the participants then could be added, by exception, to the basic agreement template. This approach avoided unnecessary boilerplate that neither party would otherwise pursue.

The Community Today: The use of CRDA development tools was prevalent among those federal agencies and departments studied. Most have drawn up standard or model CRDAs. Each agreement, however, presents different circumstantial variations. An inability to agree on CRDA negotiation items is the main reason cited that CRDA negotiations often last as much as twelve to eighteen months. This length of time is viewed universally as unacceptable with most industry timelines. The most favorable results occur where a model CRDA template exists, where continuity is assured by maintaining contact at the technology level, and especially where there are clear expectations on both sides.

Timely and Informed Negotiations

Turning to the negotiation phase of the CRDA process, a few innovative technology transfer managers and organizations stand out. It is here where all the goodwill, sound planning, and close interaction at the technology and business levels can pay off in a timely agreement, or alternatively can be lost if care is not taken. In one respect this may prove the most challenging aspect of the CRDA process, since additional functional support perspectives (e.g., the legal departments) are represented in the venture at this point. In any case, the same sensitivity to the needs of each partner, as illustrated above for the planning phase, must continue during negotiations for success to occur.

The Air Force electronics systems, command and control product center, ESC, demonstrated outstanding success through its Cooperative Technology Transfer Center manager in making the CRDA process operate within an austere budget. This visionary leader was well versed in the technology and was actively involved with the technology community inside and outside the government. Moreover, he was cognizant of the CRDA process and the enabling legislation and policy. He was able to mesh these ingredients to make the CRDA planning, negotiation and execution phases work smoothly. Specifically, a key result was that the center became adept at recognizing worthwhile technology partnerships. It also became proficient in brokering partnerships, to the point where it held its Service's record for timeliness. Success was attributed to active involvement by this manager's office in both technology and business area activities throughout the CRDA process.

Wide ranging and thoughtful recommendations (and cautions) on negotiation issues were offered by a variety of experienced industry and academic leaders to assist others in avoiding some of the pitfalls outlined earlier. An overriding factor in the message from these entrepreneurs was that the critical nature of technology cycle times must be acknowledged in a negotiation process that can support "windows of opportunity" in technology exchange.

The Community Today: Intellectual property barriers, including the persistent "FAR mentality" and adversarial distrust have been difficult for some organizations to shed. Many stories persist about a lengthy negotiation process in which a particular relationship, that had been held together by close technical level interface, was stymied at that point where bureaucratic support offices

After the technical representatives agree that their needs and capabilities are matched, it is time to get support from the legal department.

became involved in brokering the process. Conversely, good preparation, coupled with an improved understanding of the CRDA mechanism and the cultural imperatives that drive each participant's organization, have been shown to be effective in minimizing the problems encountered and in easing the negotiation process.

Execution of Successful CRDAs

Numerous excellent examples exist of CRDAs that produced the desired results. Department of Defense organizations in particular have overcome various hurdles to effectively engage in technology transfer and cooperative R&D activities. Several have found early success in sustaining their operational customers and supported weapon systems through CRDAs. Other organizations, both private and public, have succeeded by recognizing what they do best and adapting to their changing market environment. The following experiences illustrate the axiom that well trained and motivated people with the right attitude are the single most important ingredient to the success of technology transfer and cooperative R&D. Well trained staff develop world-class technical products and energize the technology transfer program.

Phillips Laboratory, whose effective program structure was discussed in Chapter One, demonstrated an innovative approach to technology exchange in its initiative to partner with private industry and a sister service production depot, Ogden Air Logistics Center to best utilize its research capabilities. In the process, it was able to satisfy its small business customer's commercial need for technology and sustain a core production capability of the depot. The three-way CRDA, although not oriented towards development of an operational weapons system, provided mutual benefits and enhanced the laboratory's mission support capabilities through technical level interchange and the preservation of core competencies.

A very successful two-year CRDA initiative by the Photonics Center of Rome Laboratories, Griffiss Air Force Base, New York, and the relationship (introduced in Chapter Four) that blossomed with a leading commercial industry company illustrated the value of choosing the right partner and nurturing a relationship.

It also represented an extraordinary outreach by a government entrepreneur with a vision for the 21st Century in information technologies to a "world class" firm, which further led to highly productive third party collaborative benefits with regional universities. This CRDA experience provided an insight from both an insider and outsider perspective into the nature of the cultural differences between the government, defense industry, commercial industry, and academia partners. The acute awareness of both principal CRDA parties to intellectual property rights, the handling of proprietary information, technology needs, and maintaining competitiveness further demonstrated how far technology exchange can go when there is commitment.

The idea of a "soft landing" in the CRDA experience was stressed by another entrepreneur, this one from a leading commercial firm. The meaning of this observation was that it was important to get worthwhile technology from the relationship to sustain and build from, not like much of the experience in the aerospace industry, for example, where programs spiral down and the energy is lost.

The Community Today: There was a consensus among participants that the CRDA experience (i.e., the outcome) did generally meet expectations. However, it was also generally agreed that it did not necessarily enhance the organization's competitive edge in any significant way, except possibly for small businesses that were able to find the perfect marriage. Most enterprises interviewed saw the relationship as enabling versus strategic, as one option versus the only option, and/or as a test of the relationship versus outcome driven. Industry's predominant view was that federal organizations do not understand industry needs, in particular time sensitivities and cycle times. Government's common view was that technology transfer was not a required mission, but rather was over and above the "primary" warfighting mission of the Department of Defense, a good idea to help the U.S. economy. The primary link with mission was to use cooperative R&D to maintain core competencies and/or preserve infrastructure from budget downsizing. The general consensus was that cooperative R&D did provide a payoff, but that relationships take time to build and time is needed to break down barriers before such enterprises become

Most firms agreed that CRDAs are enabling, rather than strategic in their impact on competitive edge. CRDAs represent one of many options for gaining access to government facilities and expertise.

productive. There was greater willingness to put more (i.e., resources, risk) into an arrangement after participants were able to see results from the test experience.

To summarize our examination of the CRDA process, cooperative research and development agreements are clearly increasing as the vehicle of choice for technology transfer. A CRDA is neither a procurement contract nor grant as defined in 31 USC 6303-6305, but is a contract in the more general sense that it is a legally enforceable document. Through CRDAs, the federal laboratories can commit resources such as personnel, facilities, or equipment (with or without reimbursement), but not funds as well as other resources as part of the agreement. CRDAs offer many benefits to the laboratories, the laboratory scientist, and the industry or university partner.

For the laboratory, the CRDA allows a flexible mechanism for transferring the results of federally funded R&D to the private sector; allows private sector participants to provide funds as well as other resources to assist in the commercialization of technology; and allows federal laboratories to receive income generated as a result of commercialization. The CRDA also affords an opportunity for federal personnel to provide expertise to private sector participants in the commercialization of their work.

For the private sector participants, the CRDA allows the nonfederal partners an opportunity to obtain rights to commercialize the results of government R&D, provides for effective leveraging of resources through a team effort, provides for access to federal expertise, and protects intellectual property for five years.

The present system has emerged as a blend of separate initiatives and experiments that supplement past policy. In the final analysis, CRDAs are an example of government leadership—not unnecessary bureaucracy. The intent of the CRDA vehicle was and remains to streamline and simplify the process of technology transfer between the federal R&D establishment and private industry.

We have now reached the point in our examination of technology transfer and cooperative R&D where we must fuel the process through the appropriate use of two very important and related catalysts for success—incentives and metrics. These subjects and the overall role of leadership in this quest for success are the topics for the next and last chapter.

chapter 6

How To Achieve Success

The preceding chapters have been instructive in pointing out the things one must do to establish a program and get a successful start in implementing technology transfer business practices: things like planning, training, negotiating, and performing in cooperative ventures. To make these lessons learned and best practices work to meet the long-term goals of the enterprise, however, the work setting itself must be supported by adequate levels of energy and commitment: things that are delivered in the ways we encourage, measure, reward, and motivate behavior.

In this chapter we examine the remaining key program ingredients that must be present for real success to occur in Technology Exchange partnerships: good metrics, the right incentives, and strong leadership. The sections below discuss the makeup and value of these ingredients to a dynamic cooperative process. The first section focuses on how and how not to measure the success of cooperative R&D. The next section suggests some appropriate incentives to

How you define success is a matter of what you measure.

[In this wheel production environment, the worker on the left measures up well if the goal is mass production of one kind of wheel. The worker on the right measures up well if the goal is to find many applications for a wheel-like object.]

enhance the technology transfer efforts of government and industry partners. The last section provides some thoughts on the role that effective leadership plays in the formula for success.

A CLOSER LOOK AT METRICS

Metrics serve as a kind of cornerstone in every cooperative venture. The metrics a manager might choose to measure the success of an enterprise depend largely on the desired outcome. For the purposes of this book, metrics may be defined as "standardized data elements that form the basis for evaluating the transfer process."[4] They allow partners in a cooperative venture to evaluate progress toward their objectives and ensure precise accounting.

Sufficient and relevant incentives provide the motivation to cooperate, communicate, and excel.

Metrics also drive behavior. Conventional wisdom suggests that most workers (and this includes technology managers and inventors) will strive to excel in areas which are directly measured, sometimes at the expense of more important activities which are not measured. Sufficient and relevant incentives create a self-sustaining momentum, motivating people, companies, and laboratories to cooperate, communicate, and excel. Metrics, because they create incentives, can propel an organization toward its goals.

The selection of appropriate metrics and incentives can be quite difficult. The results of our research leading to this book concluded that a significant number of technology managers believe their present incentive systems need to be revised (Figure 6-1). Technology experts charged with cooperative R&D ventures must act with vigor. Each individual must be offered an incentive to make technology transfer a success.

To better understand the utility of metrics, we will examine three characteristics in detail: category, focus, and scope. As Figure 6-2 shows, these characteristics indicate whether the metric pertains to an activity or an output, evaluates an organization or an individual, and

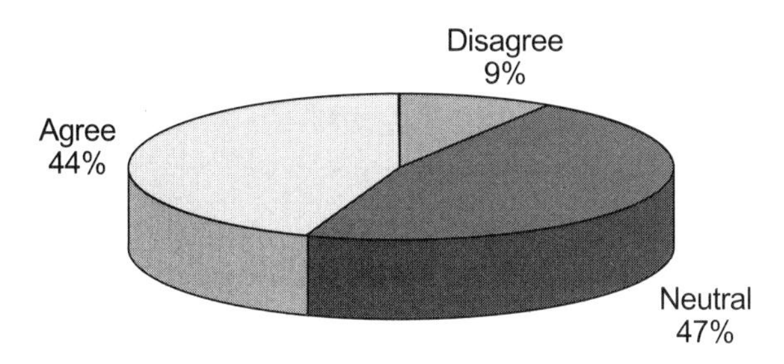

FIGURE 6-1. Responses to the survey statement "The Current Reward and Incentive System Needs to be Changed."

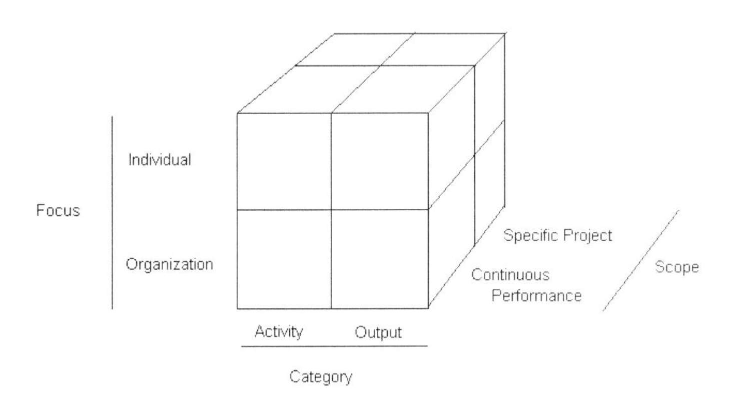

FIGURE 6-2. Description of metrics.

evaluates one specific and finite project or the continuous performance of an organization.

Category of Metric

A metric can measure two different aspects of a process: activities or outputs.[5] An activity metric measures the amount of effort applied to a venture. It observes the means, not the ends. An organization might measure the number of CRDAs signed or the number patent applications submitted. The activity metric may also be reported in units of activity per person. For example, a metric might be the average number of hours devoted to technology transfer per person in the organization. An activity metric considers neither efficiency nor timeliness. Moreover, it ignores the quality, cost, and quantity of the final product.

An activity metric has the benefit of measuring, ideally *in real-time*, the amount of work that organizations and individuals invest in cooperative ventures. This metric is capable of continuously evaluating trends instead of simply providing snap-shots. Because such an activity metric does not judge results, risk-taking is not penalized. Using an activity metric, a project that produced non-viable inventions could receive an identical evaluation as a project that produced the next generation of microwave devices.

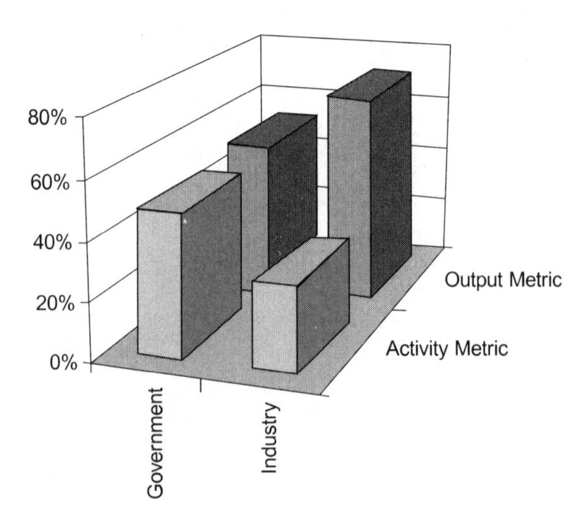

FIGURE 6-3. Type of metric employed by sector.

An output metric measures ends without consideration of means. It might consider the number of patents issued, jobs created, or products produced. The survey conducted by the Technology Exchange research team found that both government and industry respondents favored this type of results-oriented focus.

The output metric may seem simple and ideal, but it carries a number of subtle flaws. Focusing on the final product incorporates neither the importance nor contribution of intermediate steps. Moreover, innovation is risky. A metric that only recognizes the final product ignores the efforts of those who labored on parallel ideas that were not successful.

Poorly crafted output metrics may deter risk-taking by forcing innovators to concentrate on areas of probable success instead of on creative ideas with unknown potential. In any given technology transfer activity, there may be a considerable time period from initial concept to final output. A metric that relies solely upon output does not produce feedback for evaluation during the project.

Poor output metrics may cause innovators to focus on the sure outcome rather than investigate potential applications.

Nonetheless, output metrics do allow for a more diligent focus on client needs. Properly designed output metrics can also measure progress toward the organization's technology transfer goals, and forge a link to its strategic objectives.

Focus on Metric

A metric can focus on two different entities: the company's performance or an individual's contribution. The former might employ the number of jobs created within the company or the amount of R&D costs avoided while the latter would use the number of hours worked or the amount of revenue generated by the employee.

It is difficult to apply a metric across these levels and measure, for example, macroeconomic effects of technology transfer. This is why in the private sector a poor metric for an individual salesman is the increase in company sales. Such a metric provides neither direct information nor causal linkage. The salesman is not convinced that he can affect the metric and so lacks any personal feedback.

Scope of Metric

Every organization attempts to measure its performance continuously in order to pinpoint trouble and better manage its resources. In regard to technology transfer, organizations might measure the number of CRDAs signed, the amount of money saved in R&D through federal laboratory contributions, cost avoidance by the federal laboratory, or the money spent for recognition of the people involved in enhancement of an application. Such metrics are often internally generated and may not be publicly released. Few companies disclose the amount of money earned from the commercialization of an invention.

Individual ventures may also have a metric to track and encourage progress. Many agreements require milestones based on this metric. For example, a project may require a certain number of laboratory hours logged or patent applications submitted by a certain date. Another metric might emanate from a monthly survey evaluation of participants.

The distinction between a continuous and a project-specific metric is not academic. Clearly, counting the number of ventures completed would be inappropriate when evaluating a specific project. Similarly, employing a metric like the number of hours committed is helpful for evaluating the project, but does not completely aid the organization with evaluating its technology transfer effectiveness.

METRICS ARE LINKED TO INCENTIVES

The management perspective views employees as open to rewards; capable and desirous of learning, committing, and achieving; and possessing imagination, ingenuity, and creativity. Time after time, studies have shown that people who are given the authority (empowered) to demonstrate initiative not only feel happier about their jobs but are also more productive.

The flip-side of empowerment is accountability. The employee must shoulder the responsibility for his or her own failure.[6] One respondent to our survey stated, "There must be reward for success as well as accountability for failure." This new perspective is widely accepted. Even management experts who do not fully support the use of incentives base their practices and theories on this optimistic foundation.

During a roundtable discussion hosted for this study, one senior government technology manager observed that "incentives are exceedingly important for people and institutions. The only way you know how well incentives are working—which should be [increased] and which should be thrown away—is by having accountability. If you take two words out of this discussion today, I suggest those two words should be incentives and accountability."[7]

Careful application of incentives can safe-guard against damaging competition due to self-interest.

While some management experts agree that incentives may foster damaging competition due to individual self-interest, most feel that a careful application of incentives is sufficient to safe-guard against this circumstance. For example, competition may be easily avoided by rewarding long-term instead of quarterly performance awards, or by rewarding individuals of a team equally based on the whole team's results.

Our research suggests that scientists often regard cash awards and salary increases as appropriate recognition of their contributions. In this light, rewards cannot be considered counterproductive because they reinforce improved performance instead of convincing an individual to perform contrary to his interests.

This same research demonstrates that scientists believe that money does not quell their creativity but gives them more confidence to explore.[8] The Intrinsic Motivation Principle of Creativity does not actually refute this assertion but, instead, simply claims that autonomy and responsibility contribute more to motivation than do external factors. However, "inappropriate reward systems" are damaging to creativity.[9]

One expert concluded "[t]he bottom line is simple: reward plans work when properly designed and supported."*[10] Incentives are powerful motivators; they bypass punishment and coercion, avoid regulation and

*Much of incentive theory has been developed within the commercial sector. This ensures that it is effective in meeting product and cost objectives. However, the origin of the research may cast doubt on its applicability to government organizations which are based on a more egalitarian and process-oriented system. To ensure the relevance of this chapter to government laboratories, we have included references which directly address federal facilities in general, and R&D laboratories in particular.

supervision. However, incentives, like any other powerful tool, must be harnessed and controlled in order to produce the desired effect.

Inadequate Metrics/Incentives

Whether measuring individual efforts or organizational objectives, activities or outputs, technology transfer in general or a specific project, there are myriad bad metrics. Managers are disheartened to see that their measurement demonstrates that a venture is failing. Yet, they rarely blame the measurement itself for creating that failure.

Our survey collected information on the most often employed metrics for technology transfer. It also requested an assessment of the respondent's cooperative experience. By comparing these two answers, the survey data analysis highlighted those metrics that were attributed to successful ventures and those that were correlated to failures.

Thus, this figure demonstrates that profit metrics were considered positive by a wide margin, while the remaining small percentage is fairly evenly split between neutral and negative. On the other hand, metrics based on the number of cooperative agreements repeated were equally likely to be considered positive or neutral, while metrics based on number of completed agreements outstripped the neutral rating, with no negative rating expressed at all for either metric.

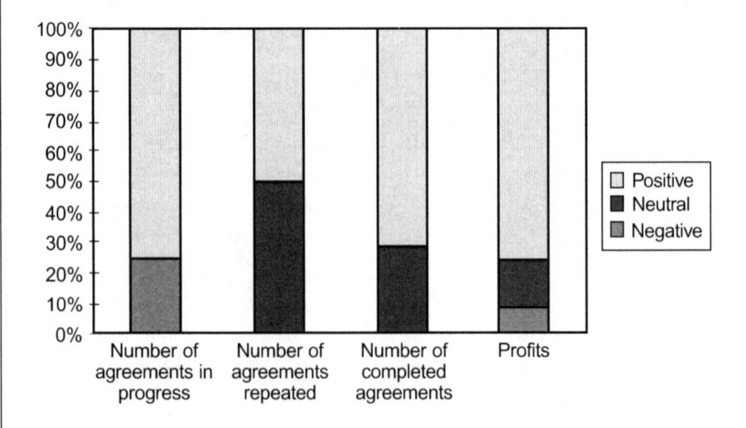

FIGURE 6-4. Rating of metric employed in technology transfer experience.

Metrics based on the number of agreements in progress were considered positive by a wide margin over negative, with no neutral rating expressed at all.

Self-Defeating Metrics of Productivity

Some cooperative agreements require metrics that are so cumbersome that they undermine technology innovation. For example, one CRDA required that every fifteen-minute block of time for every technology expert be accounted for in an attempt to optimize productivity and focus. The company that was burdened by this requirement found that the bookkeeping alone was enough to undermine exactly what they were trying to measure.[11]

The CRDA Count

The CRDA count has been the topic of much discussion and even those who employ the metric admit that it is less than optimal. Figure 6-5 depicts the responses by sector from the Technology Exchange survey. The number of government respondents who selected this metric was more than twice that of industry respondents.

This metric is irrelevant to strategic objectives and personal goals. It does not gauge larger effects on the economy: it measures neither job creation nor increases in the GNP and it does not encourage notions of ownership or accountability. As one survey respondent stated, "Merely counting the number of agreements signed fails to recognize the difficulty . . . of bringing a project to a usable conclusion."

In terms of incentive, this metric is often worse than just unnecessary; it may even be destructive. For example, tracking CRDA counts could prompt technology managers to scramble to sign as many agreements as possible without regard for the quality or result of the agreements.

Furthermore, a CRDA count is an example of a low-level count which simply creates new data while ignoring the larger impact of cooperative R&D.[12] Thus, it reduces the utility of the metric, clogs upper-level management with poor data, and invites "distracting and debilitating micro-management."[13] It is more meaningful to the government technology transfer manager

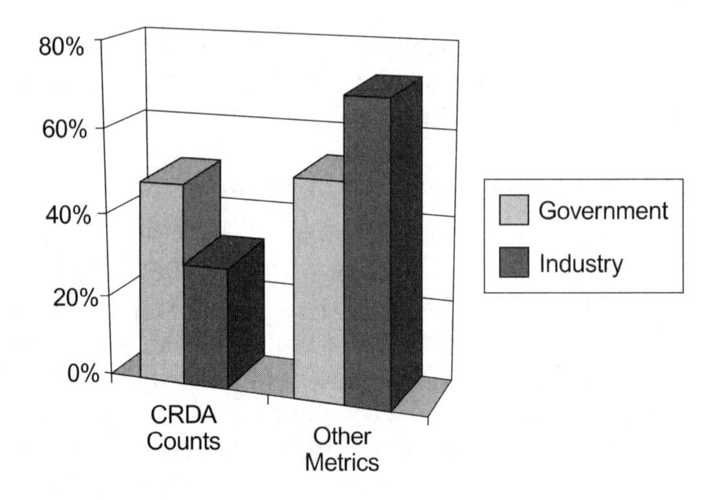

FIGURE 6-5. Respondents who used "The Number of CRDAs Completed" as metric.

whose job it is to be a facilitator, than it is to the industry or government laboratory R&D manager. It is very meaningful to the laboratory just getting its first few CRDAs and learning the process, and it is used as a first-order measurement of workload to the ORTA or supporting organization. It happens to be the only accurate measurement available in many cases.

Meaningless Metrics Behind The Funding Gap

Contemporary R&D programs often earmark funds for the initial product design and final manufacture, but not for the intermediate prototype construction. This omission, known as the "Funding Gap," forms a bottleneck which leaves many blueprints untouched on the shelves and the resources that went into their development wasted. In order to more consistently and evenly fund the R&D process from start to finish, the incentive must focus on the final product, not a stage in between.*

*Roundtables One and Two were held at the Army Navy Club (hosted by the Technology Exchange research team) in late June 1994 in Washington, DC. Proceedings of these discussions were limited to tightly restricted distribution.

For example, a metric that is based on numbers of prototypes will undoubtedly produce a large number of prototype designs. However, there will be little regard in the designs for subsequent quality, price, or marketability of the final product. The lesson we can learn here is crucial. Creation of a viable product depends on consistent funding. Metrics which isolate the subtasks of innovation, design, and manufacture undercut a consistent funding approach.

John Preston has graphically depicted the approaches to funding taken by corporations in pursuing technology development. The approach represented by the curve labeled A in Figure 6-6 shows a small amount of funding extended over a long period of time. This allows the company to post a smaller quarterly loss for R&D investments. The opposite approach is represented by the curves labeled C, where large amounts are invested over a short period of time. The curves are truncated to indicate that the company (eventually) switches over to an A curve to minimize losses. The optimum curve is that labeled B. This approach represents a high initial investment, but it produces a high rate of return in a short time once sales begin. Adopting a B curve funding strategy is the best way to effectively manage diminishing resources.

Today's pace of commercial product development calls for an R&D investment strategy that will overcome the "funding gap" and rapidly apply an innovation.

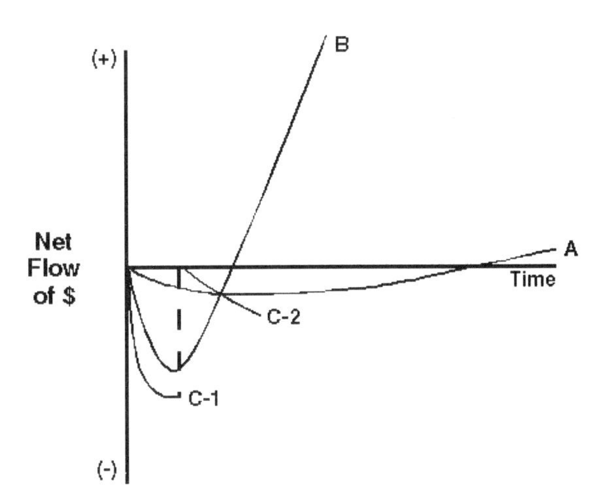

FIGURE 6-6. Corporate approaches to funding technology department. (Source: Mr. John Preston, MIT)

Personal Awards

The United States Air Force requires that cooperative ventures in which it participates "recognize and reward the people responsible for the transferring technology."[14] Technology experts who provide "exceptional value" to a cooperative venture often receive cash awards from their sponsoring organization. The recipients are often selected by the head of their agency.

In many cases, such attempts to provide a personal award are insufficient. They lack the objectivity necessary to generate wide-spread enthusiasm. Because the award is filtered through leadership, an expert who contributed significantly might be inadvertently neglected.

These few examples demonstrate that metrics are, indeed, powerful incentives. Perhaps the best evidence for this assertion is the fact that bad metrics produce unproductive behavior. Metrics, good and bad, cannot be neglected. They exist. Only by recognizing their potency can they be redesigned to promote success.

WHAT METRICS/INCENTIVES WORK BEST

Based on the results of our study, roundtable discussions, and surveys, as well as outside research, we have developed several optimal formulas for metrics which achieve success through appropriate incentives. This section will describe these in detail, along with their advantages and disadvantages.

The optimal metric and incentive system involves two different but concurrently operating categories of metrics with a minimum of obtrusive measurement and a concentration on timeliness, cooperation, and the final product. One metric is an intermediate indicator focused on activity for in-house, real-time guidance; the other is a formal output metric on which many cash rewards are given and strategic decisions are based.

Intermediate Indicators

We suggest the introduction of a new class of metric that is less quantitative and less formal than previous types. An intermediate indicator* allows venture part-

*This category of metrics originally was suggested by Berger.[5] Her explanations provide helpful further reading.

ners to reap the benefits of many different types of metrics without incurring many of the costs. This indicator would not be used to formally evaluate the success of the venture. Instead, it should be informally, internally collected and analyzed in order to provide managers with real-time feedback on workers' performance. It should focus on activities for project progress.

An intermediate indicator should be used to reward team performance, thereby avoiding the dangers of individual rivalry. By granting promotions and individual recognition based on publicized criteria and this intermediate indicator, the organization is rewarding cooperative achievement and hard work. Few if any cash awards or bonuses should be based upon this indicator to reduce the stigma of cut-throat busy work.

Formal Metric

The formal, externally publicized metric for cooperative ventures should measure three things:

- time to final production (measured in days since commencement)
- estimated revenue generated from the final product
- progress toward technical specifications.

The assessment of the project, the assessment of technology transfer in general, and individuals' incentive pay should be based upon this formal metric.

Intermediate Indicator	Formal Metric
Measures Activity ("Means")	Measures Output ("Ends")
Focuses on the Individual	Focuses on the Organization/ Team
Provides Real-time Evaluation of a Specific Project	Provides Post-closure Evaluation of a Specific Project
	Provides an Evaluation of the Organization's Technology Transfer Program

FIGURE 6-7. Optimal metric structure.

For cases involving long cycle times, it helps to set up deadlines, metrics, and goals for each segment of the process.

If the cycle time for the product is substantial, it might be advisable to segment the development into several smaller, shorter portions each with its own deadline, metric and technical goal. However, in order to avoid the Funding Gap, segmented ventures should rely upon clear and consistent work plans which evenly incorporate all stages of product development.

Time Sensitivity

Suitable metrics must include a measure of time. This can be accomplished in one of two ways. One option is to measure the time elapsed since a deadline, e.g., the number of days that have elapsed since the negotiated deadline for the creation of a working prototype. A better metric is a measurement of the duration of an event or step in the process, e.g., the number days to the creation of a working prototype.

While timeliness is crucial, performance reviews should not be conducted in intervals of less than six months. This interval allows workers to take risks, prepare well, and avoid quick fixes. Further, it reduces accounting requirements. Some experts interviewed during our site visits and benchmarking trips recommended a metric for the duration of the negotiations. A close measurement of this often protracted process would encourage haste by publicizing delay.

Intermediate Indicator	Formal Metric
Incentive Based on Team Performance: Recognition and Responsibility	Incentive Based on Team Performance: Shared Cash Awards
Incentive Based on Individual Performance: Recognition and Promotion	Incentive Based on Individual Pay-for-Performance: Bonuses

FIGURE 6-8. Optimal incentive structure.

Tying Individual Performance to the Project Outcome

A successful metric must involve and energize the individuals engaged in technology transfer. It must allow them to benefit directly from hard work and commitment. The most successful individuals and organizations "bet the ranch" on the success of a technology. Many managers feel that providing each employee with the authority to make a project sink or swim precipitates a phenomenal increase in productivity and commitment. This most primitive of incentives, survival, may be incorporated with varying degrees of severity into metric and incentive systems.

The most radical approach to this ideal is basing individual salary or bonus and team budget directly on the formal metric. The faster the teams produce a viable, quality product, the larger their year-end bonuses and larger funding for the next project.

This profit-sharing approach must overcome the often significant delay between concept and product. This concern is minimized if the intermediate indicator is employed simultaneously. The approach is less easy to apply if individuals have been working on several projects that provided valuable information, but no specific applications. Nonetheless, tying the bonus to the success of the product is a powerful and useful incentive.

Prizes and Awards

Prizes and awards can be excellent motivational tools when handled well, yet they can have disastrous effects if handled inappropriately. Based on our research, we offer several guidelines to follow in order to ensure success.

First, the cash awards given to individuals must be contingent upon results. This further ties the individual to the organization's mission. On the other hand, awards that include recognition and promotion should be based upon activity as measured by the intermediate indicator in order to reward team work.

Second, both the cash and recognition awards must publicize clear and precise selection criteria in order to eliminate any claims of subjectivity or favoritism. Further, by highlighting those behaviors that will win an award, management is directly creating a metric and incentive for employees to perform as requested. These criteria should be based upon the appropriate intermediate indicator or formal metric.

Third, the award must be comparable to the effort that produced it.[15] A prize that is deemed insufficient will actually create a disincentive to further performance. A prize that is too large may establish a weighty and dangerous precedent.

One industry expert recommended two parallel awards: one for original invention and the other for contribution to execution.[16] The former encourages creativity in scientists for their inventions. The latter fosters commitment and problem solving in the support staff. This dual award structure is especially germane to technology transfer, where the two activities must be but are not always mutually dependent.

Bridging the Communications Gap with Incentives

In Chapter 2, we introduced the notion of a culture chasm separating the public and private sectors. As lines of effective cross-cultural communication are built to overcome this barrier, government and industry strategic objectives will also begin to converge upon a common set of goals. When the gap is finally bridged, a suitable incentive system must address and encompass both sets of goals.

There are several different vehicles for technology transfer which already rely upon traditional incentives. These vehicles rest on principles of market forces and incorporate motivating metrics. However, there are some slight amendments to these vehicles which could make them even more powerful.

Cost Sharing

Modified cost-sharing arose as a popular vehicle to encourage commercial companies to work with the federal laboratories. It is better than the traditional grant because it forces the company to invest in its own success.

However, the incentives built into traditional cost-share agreements might be undermined by a partner's ambivalence. The organization might be unwilling to "bet the ranch" on a specific technology because it is not convinced of the product's commercial viability. Hesitation and lack of commitment leave the technology on the drawing boards, regardless of its commercial potential.

A solution to these problems is cost-sharing based on progress toward production. The schedule for reimbursement by government to industry for previously incurred costs might be linked to timely progress toward product release. For example, the percentage of reimbursed money might be based on product sales by a certain date. Failure by a company to adequately develop a technology will eliminate the reimbursement altogether. Failure to meet the deadline might also reduce the size of the reimbursement. This incentive structure would create "passionate behavior" within industry to produce a marketable product of good quality in a timely fashion.

We have shown the link between metrics and incentives. Technology managers should be aware that their metrics not only evaluate progress but also drive behavior. Incentives will provide the enthusiasm and commitment required to increase performance. However, these incentives and the metrics on which they rely must be selected with care. The use of a dual system of metrics creates powerful and well-directed incentives. An in-house, intermediate indicator focuses on real-time activity and team performance. It is the basis for team and individual recognition. A formal output metric measures timely progress toward technically adequate and commercially viable results. This metric is the basis for cash bonuses and an evaluation of the organization's performance in technology transfer.

Technology transfer has much to offer the laboratories and companies involved. However, the people engaged in cooperative R&D must be energized. Some even claim that our national competitiveness depends upon successful technology transfer. Metrics and the incentives they produce are keys to this reinvigoration, but the implementation of these powerful tools does not just happen. To become a strong motivating force for institutional change they must be crafted thoughtfully, advocated clearly, and, most of all, be relevant to what the organization wants to ultimately achieve. It is in creating the right organizational and programmatic climate for an energized workforce to perform where the role of leadership is pivotal.

THE ROLE OF LEADERSHIP

Incentives are powerful tools. When selected and implemented with care, incentives can contribute greatly to success; however, if inappropriately applied, they can create disaster. Incentives, in the form of metrics, exist in

When selected and implemented with care, incentives can contribute to success; if inappropriately applied, they can create disaster.

every organization through an implicit—and sometimes explicit—drive to accomplish whatever is measured. In order to harness the power of incentives and focus employees constructively toward organizational objectives, managers must also analyze and adjust their metrics. Herein lies an important role of leadership.

Ensuring Technology Transfer is a Viable Concept

As we noted in Chapter 3, technology transfer is a young but maturing process. Some of the first studies on technology transfer successes are being used to improve or guide others as they enter into partnerships with large, medium, and small firms in cooperative R&D partnerships. Some have argued that the federal program is flawed, and are trying to evaluate the impact of technology transfer activities to prove or disprove their validity. Their findings have indicated that sometimes the benefits may be long term or even intangible, and often the benefits defy quantitative measurement. These results do not suggest that technology transfer should be abandoned, nor do they support a decision to ignore the potential of technology transfer as an investment strategy. However, the preliminary data do indicate the need to prioritize cooperative R&D alternatives within the framework of an overall (master) technology process and against "real" war-fighting mission requirements.

Others assert that the cultural gap is too large and that the costs of building enabling bridges between the two sectors outweigh the benefits to be derived. Certainly scarce resources, both people and money, are needed to build a good program. Likewise, the challenges are many. However, if we have not yet measured the real benefits of technology transfer at the technology or capability level, then it would certainly be premature to judge that it is not worthwhile.

The program is flawed only if we stop here and shy away from an incremental approach that will succeed. It is true only if future business practices by government organizations are not able to adapt to new realities in mission support. It is true only if the private sector, i.e., the marketplace, cannot use the "world class" technology and resources within the federal establishment. This means that the full power of the American economy must be factored into national solutions, including security ones, in order for the U.S. to remain a leader. It also means that the timing is right for good leadership that can make a difference in how well we perform this task.

Building Upon the Accomplishments of Strong Leadership

We have observed and suggested through case studies that those organizations with a clear corporate strategy and an educated and empowered workforce can and have made technology transfer and cooperative R&D work for their organization. We've also seen that much is being done from a legislative and policy standpoint to enable the success to occur. The overall implementation, however, has been sporadic and limited to those clearly out front entrepreneurial organizations, both public and private, that knew what they wanted and worked to achieve their objectives.

In the following subsections we review where the technology transfer community is today. To help do this, we'll relate to some of those best practices discussed earlier and a few others observed that will demonstrate the clear role that an enlightened leadership has played. These corporate level initiatives are but a few of the many examples available where visionary leadership, both inside and outside of government, is taking place. The need for total commitment from the top down, more than mere advocacy, is essential to achieve the necessary cultural change. In a sense these organizations and their efforts have helped pave the way for others.

Total commitment from the top down is needed to achieve the change that will remove the cultural barriers to successful technology transfer.

Joint Strategic Planning

We have started to build a coherent national strategy with enabling policy and legislation. We have seen how recent government policy directives more clearly acknowledge the two-way nature of technology transfer activity and place more emphasis on cooperative R&D missions. Likewise, should the Department of Defense establish a funding program element for technology transfer, this would set the stage for real progress. At the national level, we must continue to plan our joint domestic investment strategy to resource the required production capabilities to accompany the development of "world class" technologies.

Vision

We have developed an awareness of what the future holds in technology for the industry and government in a post Cold War security environment and an increasingly competitive global marketplace. To a limited extent, there has been a recognition of "crisis" that facilitates progress in charting a new course and fosters the cultural change necessary to institutionalize it.

Integrated Investment Strategy, Program Structure and Focus

Too little government progress has been made to develop a coherent mission-enabling and fully integrated technology investment strategy. Each organization could benefit from a planning model such as the one offered in Chapter One to government organizations for structuring a technology transfer and cooperative R&D program. As you recall, it begins with a matrix of laboratory technologies and expertise available, followed by a validation of those considered useful by industry, and then a strategic investment analysis to provide a multigeneration and multipurpose road map of industry needs. Again, it is important to remember that whatever program is put into place, it must add value to the organization.

Bridging Cultural Differences

Methodical and plodding progress has been made in somewhat reluctant and linear, rather than parallel fashion so far. There have been some real success stories to date, notably in some of the training programs illustrated in Chapter Four that have managed to make technology transfer part of everyone's job. There has been almost universal agreement that progress here begins with the attitude of those involved.

People and the CRDA Process

Technology transfer is a contact sport, not in an adversarial context, but in close, face-to-face, and coordinated working level relationships. This theme was common to virtually every successful venture studied. The value of this interface played a critical role throughout the CRDA process as illustrated in the numerous examples included in Chapter Five.

One government laboratory has demonstrated that it is possible to structure a program that puts technology transfer on the personal level. Recognizing that entrepreneurship is notably lacking in a federal laboratory environment, this particular laboratory encourages employees to develop their ideas on commercializing a product by offering a sabbatical with benefits for up to one year. There is no obligation to return if they meet with success in the business world, nor is there an obstacle to returning if they are less than successful. This linkage of a sabbatical to commercial development opportunity signals a change in mindset that encourages the people in the laboratory to think in new directions.

Metrics

Considerable effort has been expended on the development of suitable metrics, notably by the Air Force, but the search for an optimum set of metrics to assess the real value of technology transfer activities continues. Measurements are generally input or activity oriented within the government, while intermediate or output type metrics are more prevalent in industry.

Incentives

Incentives have been built into enabling legislation and policy, but to become more effective, they must be more closely tied to performance metrics and to strategic output objectives. For example, employee performance review criteria for technology transfer activity have yet to be fully implemented for virtually all federal participants in our study. On the other hand, bold initiatives such as an entrepreneurial leave concept, developed by a Department of Energy laboratory, have encouraged technology excellence, while providing some employment return rights (i.e., a parachute) to those willing to venture into potentially lucrative commercial technology partnerships.

Wariness Persists

Today's economic realities may force us to decide and act before long-term evaluations are completed. Early results from technology transfer successes indicate we are on the right course. However, the consensus in the community today is that the jury is still out on whether dual use and technology transfer have aided in achieving long-term goals. There has been some enthusiasm for policy initiatives and programs. Most, however, do not yet view these as part of a long-term planning thrust to integrate into mission capabilities or to leverage investment resources in this way. Technologies in photonics, information, and environmental systems, composite materials, etc. that complement core activities are seen as critical for future mission support and as commercial growth areas.

PRESCRIPTION FOR SUCCESS

Policy mechanisms being put in place support a vision and strategy of technology transfer with, not to, industry, and a focus on the customer as the user of the application.

Any prescription for success must promote a management philosophy, people policies, and an organizational structure so that technology decision making is diffused downward (not upwards as has been the trend). This includes funding and manpower authority, but with accompanying accountability, to allow organizations to leverage technology and infrastructure resources. The policy mechanisms being put in place with the latest legislation reflect the need for a corporate level vision and strategy of technology transfer with (not to) industry. For government this means working with industry, academia in helping mission supportability in more ways than just cheaper products. For ultimate success, the vision and strategy adopted must focus on the customer as the user of output (as opposed to the one who has the money).

Some concluding advice for the would be participant, private or public, is given below.

1. Pursue market outreach that is targeted to a worthwhile or useful product (technology), realizing that the product is paramount—i.e., make sure you have something someone else wants.
2. Keep the legal and other staff agencies in the loop, but keep the primary interface at the technical level and out of the control of these supporting activities at least until clear goals, objectives, and mutual expectations are worked out.
3. Seek a standard simple (CRDA) template, perhaps one that has been jointly developed, that fits most situations, but is flexible enough to address specifically expressed needs.

4. Select and train good people for a "lean and mean" technology transfer program, with both a working background in technology and a business and marketing experience represented.
5. Seek (or provide if in a leadership position) top-down support and advocacy for a strategic planning concept that fully integrates cooperative R&D and technology transfer investment planning into the organization's technology master process.
6. Train and indoctrinate people on cultural differences, and in particular the need to meet each other's expectations for a positive outcome and to overcome the barriers that prevent it.
7. Exploit the Internet and World Wide Web to identify partnering opportunities and publish accounts of "best practices" and "lessons learned."

Finally, get started now—it is the most difficult step, but one that you can take. The potential for successful technology exchange makes it worthwhile for your organization to actively consider now. Why wait?

Summary

We have established the importance of technology transfer as an integral part of the U.S. national security strategy (Chapter 1) and examined the common perceptions within the public and private sectors in outline and more detailed form (Chapters 2 and 3). We discussed bridging organizations and the importance of training (Chapter 4) and summarized the negotiation points, intellectual property concerns, and some of the fine points that differentiate CRDAs from procurement contracts and other agreements (Chapter 5). Finally, we focused on the need for metrics and incentives, and the role leadership plays in cultural change. Now it is time to summarize the lessons learned.

Before the fall of the Berlin Wall, the Department of Defense was committed to maintaining a separate technological and industrial base to meet the defense needs of the country. Today, with shrinking defense budgets and increasingly sophisticated commercial developments, we are faced with a need to converge our R&D efforts to leverage from the best offerings from both government and industry.

We are at a point in time where strong visionary activist leadership, both inside and outside the government, is needed. From our year-long study we have discovered or confirmed the following:

- The most innovative and active technology transfer programs are found in organizations with a "command climate" or "corporate culture" that exhibits a commitment, within limited resources, to achieving change. A spirit of advocacy is demonstrated at each level within such organizations.
- By and large, the policy mechanisms (i.e., regulations, directives, and instructions) are in place. What separates the organizations that succeed from those that "do less well" in technology transfer is attitude. This attitude is concerned with exchanging knowledge with and between partners. Organizations which do less well tend to view technology transfer as a one-way street, namely from government to industry.
- Organizations which consider themselves successful at CRDAs (whether in terms of achieving a high count of CRDAs, patent licenses issued, or royalties received from licenses and CRDAs) tend to decentralize the decision authority to identify and enter into such agreements, and they also reward the teams and individuals who participate in dual use technology development or technology transfer activities. The people involved in technology transfer recognize that their activities are an integral part of the organization's mission.
- Organizations which are satisfied with their execution of CRDAs encourage their staff to learn more about the "other side." Large and medium-sized firms actively pursue government market intelligence by sending both technical and program development staff to conferences, symposia, and advanced planning briefings to industry. Smaller firms seek out partnering agreements with larger established organizations, join SBA-sponsored seminars, or make select contacts directly through multiple attempts to locate needed expertise. Similarly, government laboratories send their staff

to business schools, arrange for sabbaticals with industry, or encourage active participation in professional societies and trade organizations.

Leadership within the technology transfer community must evolve from being merely transactional to truly transformational. The traditional reasons organizations have for forming partnerships (e.g., to share risk, bring together complementary resources, or collaborate to overcome barriers to markets) concentrate on immediate, tangible results. These relationships are transactional. For them, contracts are primarily a vehicle for creating precise, legal, and formal agreements that clearly define the rights and obligations of the involved parties.

Through CRDAs, organizations can build a superstructure on the foundations of traditional relationships, and forge longer term alliances. CRDAs offer more flexible, less legalistic, and more informal options to all parties. Successful CRDAs are transformational in the sense that they are truly based on a "win-win" proposition. All parties are enriched by the exchange of knowledge; hence, all parties are transformed.

Joseph Badaracco, author of *The Knowledge Link: How Firms Compete Through Strategic Alliances*, lists the types of knowledge that can be called upon in a teaming environment. He notes:

> First, there is a vast pool of potentially commercializable knowledge in the world, and it is expanding rapidly, perhaps at an accelerating pace. . . . Second, a growing number of countries, companies, universities, and other organizations are contributing to this pool of knowledge. Third, some of this knowledge is migratory. It can move very quickly and easily because it is encapsulated in formulas, designs, manuals, or books, or in pieces of machinery. . . . Fourth, some of the knowledge being created around the world is embedded knowledge, and it moves slowly. The reason is that embedded knowledge resides in relationships, usually complex social relationships. A team, a department, or a company sometimes "knows" things that none of its individual members know, and some of its knowledge cannot be fully articulated.

Embedded knowledge resides in human relationships, not the technology.

CRDAs can tap into both migratory and embedded knowledge. In fact, a successfully managed and executed CRDA is more likely to transfer the second type of knowledge due to the close partnering that results from all parties belonging to the same, integrated research team. To paraphrase Badaracco, parties entering into CRDAs are creating a knowledge link, an alliance through which both organizations learn or jointly create new knowledge and capabilities. These partnerships reflect the special character of embedded knowledge: it is sticky—it moves slowly and awkwardly among the staff of both organizations. For one organization to acquire knowledge embedded in the routines of another, it must form a complex, intimate relationship with that organization. The knowledge cannot be put into a formula or a book and then exchanged for cash.

CRDAs allow two or more parties to combine know-how and skills in a unique way. CRDAs enable this working relationship, which is highly leveraged and transformational to the individual parties. Participating in a CRDA is like competing in a three-legged race.

The three-legged race is a fixture of family picnics. It is characterized by the preliminary selection of two people who are bound together, the one's left leg tied to the other's right leg, and an awkward race to the finish line, competing against other similarly bound pairs. But the three-legged race also makes the perfect analogy to the new paradigm for technology exchange. Consider the following similarities:

- *Communication, patience, and coordination.* This activity places a premium on forging and maintaining a solid, working relationship. Each participant is accustomed to a certain style and pace. Once tied together, the partners must overcome their original customs and personal preferences to find a common gait and develop realistic expectations.

- *Contact.* The partners are in constant touch with each other. Attempting to avoid contact with a partner will only cause an embarrassing and time-consuming tumble to the turf. It is important for both sides to maintain continuity.

- *Perseverance.* Even the most skilled and experienced pairs trip and fall. But they untangle themselves and continue the race. Further, the more the partners run together, the more comfortable they become with each other, and the better they run. However, if either partner balks, the pairs trip and fall, making it difficult to finish the race.
- *Diversity.* Pairs are not necessarily formed of people of compatible sizes. Often a small person and a large person, or a tall and a short person, must cooperate and race together. The difference between them is not detrimental. Perhaps the shorter one can see potential obstacles under foot better, and the taller can see the field of competition better.
- *Awkwardness.* Although participation in this race can be smooth, easy, and immediately fulfilling, the race is not necessarily going to proceed in this fashion. The rules of the game are more likely to create awkward, cumbersome configurations. To be successful, pairs must recognize the awkwardness in order to overcome it and focus on moving to the finish line.
- *Timeliness.* The driving motivation behind the event is winning, and the prizes are sometimes desirable or even valuable. Each pair wants to be the first to cross the finish line, so competition tends to be fierce. Crossing the finish line second, like entering into the market second, can be very disappointing. The pair that overcomes the awkwardness and communicates effectively and efficiently with each other, strives to the same goal, and has an incentive to win will triumph.

Winning isn't everything. Although the competition itself is real, even the pair that finishes last has accomplished something positive. It has gained some experience in cooperation, and has benefited and enjoyed the exercise. The three-legged race is not a win at all costs proposition. Perhaps this is the reason it is such a popular and welcome activity at any family picnic.

To compete in a three-legged race, the partners must overcome personal preferences and awkward constraints, agree on a gait, and communicate with each other each step of the way. To win, the partners must negotiate the various obstacles on the course, recover from any missteps and falls, and be flexible in selecting the path to the goal. Unlike the relay race, in which the lanes are well defined and the baton is passed from one racer to the next and finally to the anchor man who crosses the finish line, the three-legged race offers many alternative routes and is over when the team has reached its goal.

Appendices

Appendix A: Excerpt from the Concept Paper *Play to Win*

Appendix B: Information on the Technology Exchange Research Team

Appendix C: Summary of Technology Transfer Legislation and Executive Orders

Appendix D: Glossary

Appendix E: Technology Transfer Mechanisms

Appendix F: Bridging Organizations

Appendix G: Model CRDA

Appendix H: Technology Transfer Related Internet Sites

Appendix A

Excerpt from the Concept Paper *Play to Win*

Technology Exchange was written in response to the AFMC recognition that domestic technology transfer and dual-use technologies are central to maintaining the Air Force's "global reach-global power" mission. The book has its origin in a concept paper entitled "Play to Win" written by Andrew Dougherty and Lowell Christy of the Economic Strategy Institute. We include the executive summary of this paper for two reasons: to offer the reader insight into our project and to recognize key industry participants in the initial study effort.

PLAY TO WIN EXECUTIVE SUMMARY

Background

In response to new global and national realities, the Air Force Materiel Command Science and Technology Directorate (AFMC/ST) enlisted the Economic Strategy Institute (ESI) to examine the research and development (R&D) investment strategies of the AMFC and selected major contractors performing R&D for the Air Force. The USAF seeks to ensure access to dual-use technologies by carefully managing the defense conversion.

Dr. Jacques Gansler, executive vice-president of The Analytical Sciences Corporation (TASC), served as the primary adviser on this study.

Problem Statement

In an era of decreasing R&D budgets and greater competition for scarce resources, how can the United States Air Force (USAF) Science and Technology Directorate guide R&D investments to gain the maximum return for the Air Force, the aerospace industry, and the nation?

Present communications between AFMC and contractors, while sufficient in the past, are neither appropriate nor adequate for the present new realities.

Process

1. Identify critical technologies.
2. Identify and recruit major aerospace contractors as study participants.
3. Conduct interviews to address the following key questions:
   - Where will industry make investments?
   - How does industry plan and invest?
   - What does industry think of AFMC's present strategy, priorities, and process?
   - What is the potential for cooperative strategic joint investment?

This "Proof of Concept" study is designed to answer these questions in a summary fashion.

Participants

ESI consulted with the chief executive officers of the following major Air Force contractors:

Allied-Signal Aerospace Company
Lawrence A. Bossidy, Chairman and CEO

Eastman Kodak
Kay R. Whitmore, Chairman and CEO

Honeywell
James J. Renier, Chairman and CEO

Loral
Bernard L. Schwartz, CEO

McDonnell Douglas
John F. McDonnell, Chairman and CEO

Motorola
George M. C. Fisher, Chairman and CEO

National Semiconductor
Gilbert F. Amelio, President and CEO

Rockwell International
Donald R. Beall, Chairman and CEO

Sundstrand
Harry C. Stonecipher, CEO

United Technologies
Robert F. Daniel, CEO

The ten candidate companies were selected because together, their work spans the scope and scale of the Air Force's R&D agenda. Each was considered representative of a class of aerospace company, and categorized by internal corporate structure and the level of research and development work performed for the Air Force. Interviews with chief executive officers, chief strategists, and directors of research and development were structured to ensure thorough and accurate sampling within each organization. All company leaders agreed enthusiastically to participate. A number of chief executive officers expressed a desire to contribute personally to the study.

Leaders from both the Air Force and industry recognized the need for a change in the process of technology development, and were extremely forthcoming with their concerns and visions—one company even disclosed to ESI its R&D strategic plan for 1991-92.

SUMMARY

On November 30, 1992, ESI reported its findings to AFMC officials, including the chief of the Science & Technology Directorate. While our assessments were not always flattering to the USAF or the companies, the report was extremely well-received as a valid set of observations. It validated the concept of "Play to Win." Following the briefings, the AFMC requested that ESI prepare a plan to expand on the observations made in the proof-of-concept study.

The AFMC asked ESI to examine in greater depth the interface issues identified in the Air Force and industry. The USAF's ultimate goal for the project is the synthesis of a new paradigm for USAF-aerospace industry research and development planning and execution. The Air Force and the private sector need to find ways to coordinate their R&D expenditures in order to most efficiently leverage the benefits of these essential technologies.

In an era of radical change, national security is no longer defined only in geopolitical-military terms. Although America's aerospace industry still leads the world, our nation's producers face increasingly hostile competition from foreign companies—many backed by foreign governments. Without a concerted effort between the US government and America's aerospace contractors, the relative international competitiveness of the nation's aerospace industrial base will continue to deteriorate.

The end of the Cold War presents an opportunity for the United States to channel its great military productive capability to commercial advantage. The United States Air Force seeks to help lead this rejuvenation of the nation's industrial base. Starting with a reexamination of research and development/acquisition, the Air Force seeks to change an historically arms-length—sometimes adversarial and hostile—relationship to one in which joint strategic planning and production is the norm.

INITIATIVES

Suggested Actions:

Convene a round-table discussion between CEOs, senior Air Force officials, and ESI at which we will prioritize the interface issues, identify promising avenues and processes to facilitate cooperative R&D planning, and set an agenda for the next stage of the study.

Organize and moderate a subsequent series of planning meetings designed to identify, refine, and develop specific mechanisms to improve communications and cooperation between the sectors

Institutionalize an ongoing cooperative approach to promote the optimal acquisition, development, and deployment of critical technologies.

Appendix B

Information on the Technology Exchange Research Team

This appendix contains information on the organizations and the individuals who participated on the Technology Exchange research team.

PROFILES OF THE ORGANIZATIONS BEHIND THE TECHNOLOGY EXCHANGE STUDY

The Economic Strategy Institute

The Economic Strategy Institute (ESI) has been created to define a new American agenda for the twenty-first century, one in which national security and success will be defined more in geo-economic than in geo-political or military terms. By challenging outmoded doctrines and examining the links between domestic and international economic policies, technological prowess, and global security issues, ESI aims to develop an integrated strategy that will halt erosion of the U.S. economic base and assure the future of America's unique promise. ESI is a non-profit organization open to all organizations and individuals that support development of such a strategy.

Write: Economic Strategy Institute
1401 H Street, N.W., Suite 750
Washington, DC 20005
Call: (202) 289-1288
Fax: (202) 289-1319

The Battelle Memorial Institute

Battelle Memorial Institute has delivered technology-based value to industry and government for more than 65 years, developing the technology behind the products of some of the most successful companies in the world. Battelle inserts technology into systems and processes to turn problems into opportunities for manufacturers, trade associations, pharmaceutical and agrochemical industries, and government agencies supporting national security, transportation, and the environment. Battelle's staff of 7,500 technical, management, and support professionals transfer technology to meet the needs of clients in more than 30 countries.

Battelle also operates three subsidiaries and maintains numerous offices throughout the United States and in select countries abroad. The number to call for more information about Battelle is (614) 424-3304 or toll free (800) 201-2011. The Battelle Home Page on the World Wide Web is http://www.battelle.org. *Technology Exchange* was produced in the Battelle Crystal City Office, 1725 Jefferson Davis Highway, Suite 600, Arlington, VA 22202. This office may be contacted at (703) 413-8866.

The Supportability Investment Decision Analysis Center (SIDAC)

SIDAC is an Air Force Information Analysis Center serving all of the Department of Defense (DoD) and its contractors. It is dedicated to the enhancement of weapon system and industrial supportability, beginning with the earliest stages of technology development or concept exploration and progressing through the final phases of the system life cycle. SIDAC is funded by the Air Force Materiel Command and operated by a team of contractors led by a Battelle Technical Services Organization. The SIDAC contract team has at its disposal the collective resources of more than 10,000 logisticians, researchers, scientists, engineers, analysts, and others involved in one way or another with the supportability mission. One significant SIDAC mission thrust area, directly related to this guide, is to promote and assist the transfer of technology developed in DoD laboratories, United States government agencies and departments, commercial activities of US persons, and other authorized users of US Government technology. The SIDAC Program Office is located at 5100 Springfield Pike, Suite 110, in Dayton, Ohio. Call (937) 254–9902 or (800) 54–SIDAC, fax (937) 254–9575, or visit the SIDAC web site at http://www.sidac.wpafb.af.mil.

TECHNOLOGY TRANSFER STUDY TEAM MEMBERS

Andrew J. Dougherty, ESI Project Director and a Senior Fellow, was Executive Assistant to the President, RFP (Research Corporation for Federal Programs) prior to joining the Economic Strategy Institute. He was founder and Chief Executive Officer of several successful small businesses in Washington, DC, Rochester, NY, and Syracuse, NY. He founded the National Security Affairs Institute and was Director, Strategic Research Group, The National War College and National Defense University, Washington, DC. After 21 years as an operational fighter pilot in the Korean and Vietnamese wars, retiring as a Colonel, Mr.

Dougherty was Executive Assistant to the Deputy Assistant Secretary of Defense in Washington, DC, from 1972 to 1975. He received his B.S. in management from the University of Nebraska in 1961 and his M.B.A. from Bradley University in 1967.

Morgan Fargarson, Research Assistant at ESI, is pursuing studies in International Economic Policy and Economics at The American University in Washington, DC. She has also worked in the Washington, DC office of Representative Bill Archer.

James Hall, Research Associate at ESI, worked previously as the trade assistant to the Chief Trade Counsel for the Senate Finance Committee (Minority) on the passage of the NAFTA. He also worked as a Legislative Assistant to Senator Bob Packwood in the areas of Commerce, Budget, and Agriculture. Mr. Hall has a B.A. in Political Science from the University of Oregon and a Diploma in European Economic Systems from the Sorbonne (Paris).

Michael Irish, Senior Research Associate at ESI, is a specialist in the areas of technology, strategy, and United States international competitiveness. While at ESI he has conducted research relating to information technology policy, the future of United States aerospace industry, and Japanese trade and industrial policy. He came to ESI with international trade experience gained in the Far East and has served with the United States Marine Corps. He received his B.A. in International Relations from San Francisco State University.

James Kelly, a Communications Specialist with Battelle, has worked on several information technology exchange projects involving government, industry, and academic subject matter experts working within a group-enabled collaborative environment called the Battelle Round Table. He is currently expanding the Battelle Crystal City Operation Internet connectivity and is actively engaged in on-line research and web page development. Mr. Kelly is pursuing an MCSE certification in Information Systems technology.

Michael D. Kull, Research Associate at ESI, is a doctoral fellow in the Management Science Department in the School of Business and Public Management at the George Washington University. His program of study is the Management of Science, Technology and Innovation. He received his Master of Arts degree from the Elliott School of International Affairs at GWU, and his B.A. in English and Writing from the University of Washington in Seattle and the University of Aberdeen in Scotland. Mr. Kull has worked for the National Science Foundation and for a number of consulting firms in international business and technology.

Ted Ladd, Research Associate at ESI, is pursuing an interest in international economics. He graduated *cum laude* from Cornell University having completed degrees in Government, Biology and Technical Sociology. His DC work experience includes the President's Council on Environmental Quality and the Committee on Foreign Affairs in the House of Representatives.

John N. Lesko, Jr., Principal Research Scientist with Battelle, assists both government and industry organizations with their R&D investment strategies and technology transfer decisions. He has written extensively on what today's leaders must "be, know, and do" to serve as effective knowledge workers. While at the Army's Materials Technology Laboratory in Watertown, Massachusetts, he managed all Cooperative Research and Development Agreements (CRDAs) and oversaw the management training of the government's technical staff. In 1994, he served as a judge for the first ever Small Business Innovative Research (SBIR) Technology of the Year Award. A guest lecturer at the Air Force Institute of Technology and the Defense Systems Management College, Mr. Lesko is a member of the

Army's Acquisition Corps and has served in the Office of the Secretary of the Army for Research, Development, and Acquisition. Mr. Lesko has earned a Bachelor of Science degree from the United States Military Academy at West Point and a Master of Science degree in Technology Strategy and Policy from Boston University.

Phillip A. Nicolai, Task Leader and Senior Research Scientist with Battelle, is Deputy Director for the Air Force Supportability Investment Decision Analysis Center (SIDAC) in Dayton. He has more than twenty seven years of successful team-oriented logistics management experience, retiring as a Colonel from the Air Force in 1992. He has an extensive background in weapon system and product management, strategic planning, and business process development as they relate to aerospace technology. Mr. Nicolai had a key role in major organizational and process restructuring activities brought on by the end of the Cold War era and the defense management review process. He earned a Bachelor of Business Administration from the University of Michigan in 1966 and an MBA from the University of Utah in 1976.

Michael L. Steve, Jr., Lead Editor at Battelle, assists in the development and delivery of draft and final edited materials for documents (including Environmental Assessments, Software Manuals, and Task Force Reports), presentations, surveys, brochures, and other forms of internal and external communication. He has served as discussion leader at conferences on publications management and information resources, and was Ohio delegate from the public to the 1991 White House Conference on Library and Information Services.

Brian Wagner, Research Assistant at ESI, received a Bachelor of Arts in Political Science from the George Washington University in May 1992. Mr. Wagner worked for the law firm of Morrison, Hecker, Kuder, Curtiss & Parish as an undergraduate and is continuing his law studies.

David Weinstien, Research Associate at ESI, graduated from Tulane University with degrees in Political Science and History. He is completing a Master of Arts degree from the School of International Service at the American University. He has worked on Capitol Hill and covered foreign and defense policy issues.

Robert I. Widder, Senior Research Scientist and Project Leader with Battelle's Advanced Technology Office, has directed studies of the technology acquisition and transfer processes of the US, USSR, and developing countries, the development of the Militarily Critical Technologies List (MCTL) and its predecessors, Emerging Technologies List and analyses of the US Strategic and Critical Materials List. He has served as an adviser on technology transfer and export control to the US Departments of Defense, State, and Commerce, and as Technical Expert to the US Delegation to COCOM. He chaired the Technical Advisory Committee on Transportation Technologies of the Department of Commerce from 1989 through 1993. Mr. Widder also is a professional engineer. He received a Bachelor of Mechanical Engineering degree from Cornell University in 1949, and has conducted graduate work in Mechanical Engineering and Engineering Management at New York University, University of Buffalo, and Bridgeport University.

Appendix C

Summary of Technology Transfer Legislation and Executive Orders

The following information was excerpted from *AFMC Technology Transfer Handbook*, Pages A-5 & A-6.

Year	Public Law (P.L.)	Name	Major Elements (Purpose)
1966	P.L. 89-554	Freedom of Information Act (FOIA)	• Provided a vehicle to inform the public about Federal Government activities • Provided the right to request agency records and have them made available promptly
1980	P.L. 96-480	Stevenson-Wydler Technology Innovation Act	• Established technology transfer as a mission of the science and engineering personnel assigned to laboratories of the Federal Government • Established ORTAs
1980	P.L. 96-517	Bayh-Dole Act	• Superseded all previous laws that give small businesses and non-profit organizations (including universities) certain rights related to inventions they developed under funding agreements with the Government (Did not give maintenance and operation (M&O) contractors right to elect title to its inventions.) • Protected descriptions of inventions from public dissemination and FOIA for reasonable period of time to file patent applications
1984	P.L. 98-620	Trademark Clarification Act	• Amended Bayh-Dole to permit M&O contractors to elect title to inventions in exceptional circumstances and national security funded technologies
1986	P.L. 99-502	Federal Technology Transfer Act	• Authorized CRDAs for Government-owned Government-operated (GOGOs) organizations • Established FLC • Provided a preference to U.S.-based business • Established technology transfer as a laboratory mission
1987	N/A	Executive Order 12591, Facilitating Access to Science and Technology	• Emphasized U.S. commitment to technology transfer • Required Government agencies to delegate authority to Government-operated laboratories to enter into cooperative agreements to the extent they are legally capable and provided authority to improve the global trade position of the United States

Year	Public Law (P.L.)	Name	Major Elements (Purpose)
1988	P.L. 100-418	Omnibus Trade and Competitiveness Act	• Mandated the establishment of regional university-based Manufacturing Technology Centers for transferring advanced manufacturing techniques to small- and medium-sized firms
1989	P.L. 101-189	National Competetiveness Technology Act (NCTTA)	• Authorized CRDAs for Government-owned Contractor-operated (GOCOs) organizations • Protects trade secret information brought into or developed under a CRDA from disclosure under FOIA
1991	P.L. 101-510	Defense Authorization Act	• Authorized federal laboratories and Federally Funded Research and Development Centers (FFRDCs) to award contracts to a partnership intermediary for services that increase the likelihood of laboratory success in joint activities with small business firms.
1991	P.L. 102-245	American Technology Preeminence Act	• Extended FLC mandate through 1996 • Allowed exchange of intellectual property between participants in a CRDA • Required a report on the advisability of CRDAs that would permit federal contribution of funds • Allowed laboratory directors to give excess equipment to educational institutions or nonprofit organizations as a gift
1992	P.L. 102-564	Small Business Technology Transfer (STTR) Act	• Established the STTR program
1996	P.L. 104-113	National Technology Transfer and Advancement Act	• Amended Stevenson-Wydler Technology Innovation Act and Federal Technology Transfer Act by creating incentives and eliminating impediments to encourage technology commercialization. • Addressed technology transfer policies in GOGO and GOCO laboratories • Assured U.S. companies that sufficient intellectual property rights will be granted to justify prompt commercialization of inventions arising from CRDAs

and that privileged confidential information will be protected when a CRDA invention is used by the government.
- Provided incentives and rewards to federal laboratory personnel who create new inventions under a CRDA.
- Clarified contributions laboratories can make under a CRDA.

Service-specific regulations and directives have also been passed. Examples include DoD 3200.12-R-4 (Domestic Technology Transfer Program Regulation) and AFPD 61-3 (Air Force Domestic Technology Transfer Policy Directive), which authorize and provide procedures for use of CRDAs.

Appendix D

Glossary

Contract. An acquisition instrument entered into between the government and a contractor for the contractor to provide supplies or services to the government. Office used to promote research and development that can subsequently be transferred to the private sector.

Copyright. Legal protection provided for original works of authorship fixed in a tangible medium of expression as provided for by Title 17 of the United States Code. Some examples of works that are copyrightable are writings, paintings, movies, sculptures, and computer software.

Cost-Shared Contract. A negotiated agreement that includes in-cash or in-kind arrangements. Patent rights are allocated as described in the contract.

CRDA (also known as CRADA). A cooperative research and development agreement as provided for by Title 15 of the United States Code. This agreement is between one or more federal laboratories and one or more nonfederal parties under which the laboratory may provide personnel, services, facilities, equipment, intellectual property, or other resources (but not funds) with or without reimbursement to the nonfederal parties. The nonfederal parties may provide funds, personnel, services facilities, equipment, or other resources to conduct specific research or development efforts that are consistent with the federal laboratory's mission.

Dual-Use Technology. A technology that has military and commercial applications.

Exchange Programs. Short-term (usually one year or less) agreements between the laboratory and others to interchange information by the unilateral or bilateral exchange of personnel. Generally, no proprietary data are exchanged, and costs are borne by the organization sending the personnel.

Exclusive Patent License. A right granted by a patent owner to one or more licensees to provide the exclusive legal right to make, use, or sell the patented invention in the United States. If several parties are licensees, then such licenses are called partially exclusive licenses.

FLC. Federal Laboratory Consortium for Technology Transfer. The organization of federal research and development laboratories and centers established pursuant to Section 10(e) of reference (b) codified in Title 15 U.S.C. Section 3710(e). It was formed to identify and mobilize the necessary resources to provide the environment, organization, and necessary technology transfer mechanisms required for the fullest possible utilization of federally sponsored research and development.

FOCI. Foreign-owned, controlled, or influenced.

FOIA. Freedom of Information Act.

FTTA. Federal Technology Transfer Act of 1986 (PL 99-502).

GOCO. Government-Owned, Contractor-Operated (facility).

GOGO. Government-Owned, Government-Operated (facility).

Intellectual Property. An intangible right that can be bought and sold, leased or rented, or otherwise transferred between parties in much the same way that rights to real property or other personal property can be transferred.

Invention. Any innovation or discovery which is or may be patentable or otherwise protected under Title 35 U.S.C., or any novel variety of plant which is or may be protected under the provision of the Plant Variety Protection Act, Title 7 U.S.C.

Laboratory. A facility or group of facilities owned, leased, or otherwise used by a federal agency, a substantial purpose of which is the performance of research, development, or engineering by employees of the federal government.

License. The contract that gives permission to make, use, or sell a patented product or process. Licensing can be exclusive or nonexclusive, for a specific field of use, for a specific geographical area, U.S. or foreign. If ownership is transferred, it is called an assignment.

NCTTA. National Competitiveness Technology Transfer Act of 1989.

Nonexclusive License. Right granted by patent holder to licensee to use, manufacture, and sell a patented article. A nonexclusive license allows the patent holder to grant additional licenses for a patented article.

NTTC. National Technology Transfer Center. The NTTC was established to serve as a gateway to a national technology transfer network, to provide education and training to leaders in business, government, and economic development circles concerning the best ways to implement technology transfer procedures, and to perform outreach services to develop working relationships and agreements with trade and professional associations and sources for technology transfer.

ORTA. Office of Research and Technology Application. ORTA is an organizational unit created under Public Law 96-480. The primary function of this office is to disseminate information on federally owned or originated products, processes, and services linking the research and development resources of the federal laboratories, and the federal government as a whole, to state and local government and the private sector.

Patent. A grant from the federal government to an inventor wherein, in exchange for the inventor providing an enabling disclosure of the invention and complying with other legal requirements, the government awards the inventor with the right to exclude other from making, using, or selling the claimed invention for a period of time usually 17 years.

Proprietary Information. Information which embodies trade secrets developed at private expense and commercial or financial information which is privileged or confidential under the Freedom of Information Act. Normally, for such information to be afforded legal protection, it should be recorded and marked as proprietary. Note: Related terms sometimes used are restricted computer software and limited rights data.

Patent Royalty. Monies or other consideration usually payable to a patent owner by a licensee of the covered patent.

RTTC. Regional Technology Transfer Center. The RTTCs, established in the six FLC regions spanning the United States, are sponsored by NASA in support of the federal technology transfer mission. The centers provide a full range of services to U.S. firms and industry within their regions, assisting clients to locate, assess, acquire, and commercialize technologies from throughout NASA and the federal R&D base.

Small Business. A business defined as a small business by the Administrator of the Small Business Administration in accordance with 15 U.S.C. 632 and implementing regulation (e.g., 13 CFR 121.3-8and 13 CFR 121.3-12). For many businesses this designation is determined by its industrial classification, the number of its employees (e.g. less than 500), and its competitive relationship to other businesses in the industry.

SBIR. The Small Business Innovation Research (SBIR) program is federally funded to promote small business participation in government programs. Characteristics of the arrangement include a 2-year confidentiality limit on data, ability of the contractor to acquire title to inventions, and multiple funding phases from feasibility to private commercialization.

STTR. The Small Business Technology Transfer (STTR) program requires five federal agencies to fund cooperative research and development projects involving a small company and a research at a university, federally funded research and development center, or nonprofit institution.

Technology Transfer. The process by which technology, knowledge, and information developed in one organization, one area, or for one purpose applied or used in another organization, in another area, or for another purpose.

Title. An intangible legal right of ownership to property

Trademark. Establishes a unique expression to identify the source of goods or services for commercial purposes. Trademark registration can be obtained from the federal government and often from state governments.

Trade Secret. Provides the right for withholding any commercial formula, device, pattern, process or information that affords a business person an advantage over others who do not know it.

TRP. Technology Reinvestment Program. A federally-funded program to help U.S. companies convert defense-related technologies to commercial applications or to combined civilian-military use.

User Facilities. Unique, complex, experimental, scientific facilities, equipment, software, and collection of expertise at a government laboratory specifically designated by the cognizant agency for use by the technical community, including universities, industries, other laboratories, and other government entities.

Waiver. The relinquishment of a legal right to do something. In technology transfer agreements, this right is usually the right of the government to obtain title to inventions developed under an agreement.

Appendix E

Technology Transfer Mechanisms

WORKING WITH THE FEDERAL GOVERNMENT

Source: "Partnering in Technology with the Federal Government: A Quick Reference"

The federal laboratories, and the agencies who manage them, are looking to industry to transfer federal technologies and expertise for commercial applications that will improve the US economy. This reference lists characteristics and features of ways to access the knowledge and expertise of the federal laboratory system and work with the federal government.

Collegial interchange, conference, publication, databases, and technical information centers. The informal and free exchange of information among colleagues

- Often becomes a precursor to long-term interactions between a company and a federal R&D laboratory or center

- Includes: (1) presentation at professional and technical conferences, (2) publication in professional journals & magazines, and (3) identification of licensable technologies, software and research-in-progress through commercial and government databases and technical information centers.
- Parties sometimes use non-disclosure agreements prior to receiving or giving out proprietary information.

Co-funded and Cost-shared Contract. A contract entered into between the government and a non-federal party in which costs associated with the work are shared as specified in the contract

- Includes in-cash or in-kind arrangements
- Must be of mutual benefit to industry and government
- Commercially valuable data may be protected for a limited period of time
- Advance waivers frequently not granted unless the contractor shares at least 20% of the total contract cost

Contract. An acquisition instrument entered into between the government and a contractor for the contractor to provide supplies or services to the government

- Can be used to fund R&D that may be transferred to the private sector
- Allocation of patent rights determined by the type of contractor performing the work—i.e., large businesses may frequently obtain a waiver on inventions where waiver criteria are satisfied; nonprofit organizations/small businesses may obtain title to inventions under the Patent and Trademark Amendments Act of 1980 (PL 96-517)

Cooperative Research and Development Agreement (CRDA, also known as CRADA). An agreement between one or more federal laboratories and one or more non-federal parties under which the government, through its laboratories, provides personnel, facilities, equipment, or other resources with or without reimbursement (but no federal funds may flow to non-federal parties)

- Although the collaboration involves the expenditure of federal funds and the use of federal personnel, services, equipment, intellectual property or other resources, no federal funds may flow to the CRDA partner
- The non-federal parties may provide funds, personnel, services, facilities, equipment, or other resources to conduct specific research or development efforts that are consistent with the laboratory's mission
- CRDA is not a procurement contract, grant or cooperative agreement as those terms are used in 31 USC 6303-6305
- Rights to inventions and other intellectual property are negotiated as part of the CRDA agreement

- Certain generated information (of non-federal party or laboratory) that qualifies as protected CRDA information may be withheld from public dissemination for a period of time up to 5 years
- Special consideration is given to small businesses and consortia involving small businesses
- Preference is given to businesses that are located in the United States and agree to manufacture substantially in the United States products that embody inventions developed under the CRDA or are produced using inventions developed under the CRDA
- US government always retains a nonexclusive or nontransferable, irrevocable, and paid-up license to practice any invention developed under a CRDA for governmental purposes
- Federal departments and agencies are developing their own policy guidelines — participants need to check with each department or agency to learn of unique requirements which may apply

Grant and Cooperative Agreement (Assistance Instruments). An agreement between the government and a recipient whereby money or property is transferred to the recipient to support or stimulate research

- Only the government can enter into these agreements; laboratories cannot

Licensing from the government or laboratory/facility to the private sector. The transfer of less-than-ownership rights in intellectual property to a third party, to permit the third party to use intellectual property

- Can be exclusive or non-exclusive, for a specific field of use, geographical area, or US or foreign usage
- Potential licensee must present plans and intentions to commercialize the invention
- Follows General Services Administration (GSA) issued licensing regulations with preference for non-exclusive license being granted. Where exclusive license is requested, requires a notice of availability, an opportunity for the public to object, and plans and intentions to commercialize
- Government obtains a non-exclusive, royalty-free world-wide license to the invention

Personnel Exchange. A transfer of personnel either to the laboratory/facility from another party or from the laboratory/facility to another party to exchange expertise and information

- Generally no proprietary data are exchanged
- Cost is paid by the organization sending the personnel
- Programs are short-term (usually one year)

Small Business Innovation Research (SBIR) Program. Federally funded to promote small business participation in government programs

- Data have a 2-year confidentiality limit
- Small business acquires title to inventions
- Provides multiple funding phases from feasibility to private commercialization

Technical assistance provided by laboratory/faculty personnel. Provided to a private sector party by laboratory/facility personnel to further technology transfer

- Laboratory must approve of the laboratory/ facility personnel consulting arrangement
- Conflict of interest on the part of the laboratory staff must be avoided

Use of Facilities. Unique, complex, experimental scientific facilities, including equipment and expertise at a government laboratory, designated by the government for use by the technical community, universities, industry, other laboratories, and other government facilities

- Includes designated user facilities and other user resources
- Research may be conducted on a proprietary or non-proprietary basis
- Full-cost recovery is required for proprietary R&D
- Class patent waiver granted in which title goes to the user and the user's proprietary data can be protected
- For non-proprietary R&D, title to inventions goes to the user but data generated are freely available
- If funded under another government contractor or international agreement, users are subject to those intellectual property clauses

These contracting arrangements offer several opportunities to the private sector to play an integral part in developing new technology. They also provide government laboratories and facilities with access to needed capabilities available only from private industry and universities, and they serve as a vehicle for communication among the various participants.

DESCRIPTION AND ASSESSMENT OF TRANSFER MECHANISMS

Mechanism	Who Profits	Who Works	Who Pays $
Informal Exchange	Both	Both	Neither
Consulting by Laboratory	Business	Government	Sometimes Business
Use of Government Facilities	Business	Both	Sometimes Business
Work for Other	Government	Government	Business
Personnel Exchange	Both	Either or both	Neither
Standard Contract	Government	Business	Government
Cooperative Agreement	Both	Both	Both
Grant*	Nonprofit	Nonprofit	Government
SBIR**	Government Small Business	Small Business	Government
STTR**	Nonprofit Small Business	Nonprofit Small Business	Government
CRDA	Both	Both	Both
Space Act Agreement	Both	Both	Sometimes Business
License	Business	Both	Sometimes Business

*Grants can only be given to nonprofit organizations.

** SBIR and STTR grants can only be given to small businesses (under 500 employees).

Information in this table was derived from *The Hunt for Technology: 10 Steps to Federal Technology*, published by the Federal Laboratory Consortium.

Appendix F

Bridging Organizations

As mentioned in Chapter 4, there are many bridging organizations. Some cater to the private sector and act as clearing houses and resources directories for the facilities and expertise of federal laboratories. Others simply fund commercial innovations. Still others link government and industry, requiring inputs from both sectors. This section of the guide offers a snap-shot of the most popular and useful bridging organizations. Each description includes an overview of the program, the most prominent resources and services, and the relevant contact numbers.

For those readers who require a more detailed analysis of the universe of domestic bridging organizations, the publisher of this guide, Battelle Press, also produces a comprehensive profile of state and federal cooperative technology programs entitled *Partnership: A Compendium of State and Federal Technology Programs*. This directory clearly describes and locates every state program and summarizes ten federal programs.

THE FEDERAL LABORATORY CONSORTIUM FOR TECHNOLOGY TRANSFER (FLC)

The office of C. Dan Brand, Chairman of the FLC, manages a network of 600 research labs and centers from 16 federal departments and agencies founded in 1974 to accelerate the commercialization of federal R&D by forging partnerships with technology users from the private sector, state, and local governments. The FLC lab locator network matches potential collaborators with laboratory resources.

The FLC connects federal laboratory members with potential users of government-developed technology. The FLC's function is twofold: first, to train technology transfer focal points in industry and government on request; and second, to reach out to industry through standing committees, brochures, exhibits, press releases, and its database, the *FLC Locator Network*. The optimal method to utilize the FLC's resources is to complete its *Technical Request Form*.

The strength of the *Locator* network lies in its ability to put the potential user in direct contact with federal lab personnel who have specific expertise in the technology area. A request for a match requires approximately two weeks to fulfill. Once the link between the two partners has been made, however, the FLC does not participate in the technical exchange.

The FLC is subdivided by region: Far West, Midwest, Northeast, Mid-Atlantic, Washington DC, Southeast, and the Mid-continent. Each region has its own regional and deputy coordinators. In addition, each region publishes directories of federal lab contacts and FLC lab representatives entitled *Tapping Federal Resources*. These directories also contain the mission statement, background, facilities/resources, areas of expertise, specific accomplishments in technology transfer, and major programs for all the FLC member labs in that region.

For more information about the FLC, contact:

Dr. Andrew Cowan
FLC Locator
224 W. Washington, Suite 3
P.O. Box 545
Sequim, WA 98382
Phone: (206) 683-1005
Fax: (206) 683-6654
www.zyn.com/flc/

NATIONAL TECHNOLOGY TRANSFER CENTER (NTTC)

The office of David Moran, Executive Director of NTTC, promotes collaborations between US companies and federal laboratories to develop and commercialize technologies. NTTC database is a NASA-sponsored information center.

The NTTC connects industry, government and academic institutions with each other in order to create suitable and mutually beneficial opportunities for technology transfer. All services are free of charge, confidential, and without further obligation. The NTTC bases its references on a substantial database which includes the facilities, technical specialties, and research objectives of over 700 federal laboratories, several universities, and many electronic and optically scanned research guides.

There are two possible avenues by which to approach NTTC. Both are simple and efficient. First, an interested party may call (800) 678-6882 to arrange a free interview with one of NTTC's technology agents. These experts are versed in both technology (many with engineering or medical degrees) and business backgrounds. The agents assess the technical problem or desire and provide a list of federal technology transfer specialists. NTTC agents estimate that it takes two weeks to orchestrate a match between industry and government participants.

The second avenue is an electronic bulletin board providing free, unrestricted, and direct access to federal laboratory capabilities. The on-line service, *Business Gold*, contains sources of federally funded research, patents for licensing, SBIR solicitations (see below), conferences, and other technology transfer opportunities. Industries may explore the bulletin board attempting to "pull" technologies from federal labs. Alternatively, federal laboratories may advertise their expertise in an effort to "push" technology into commercial markets.

While NTTC technology agents direct interested companies to technology specialists within federal laboratories, they will not provide any further assistance, neither recommending optimal vehicles nor describing historical successes. In addition, the NTTC estimates its database contains just seven percent of the available literature. This number is growing rapidly as the two year-old service matures, but it still does not represent the wealth of material that is available.

For more information about the NTTC, contact:

Hot-line: (800) 678-NTTC
Marketing Division: (304) 243-4440
www.nttc.edu

NATIONAL INSTITUTE OF STANDARDS AND TECHNOLOGY (NIST)

The National Institute of Standards and Technology, a division of the Department of Commerce, performs cooperative research with various companies. NIST does not disperse funds. NIST's primary publication, *Guide to NIST* (SP858), is a directory of programs and contacts of approximately 100 pages. Press releases and information on NIST-sponsored activities are available through its free mailing list, which contains approximately 25,000 organizations. NIST also sponsors seminars, standing committees, professional societies, and peer-to-peer communication.

While NIST will respond to requests, it does not actively pursue industry partners. If a potential industry partner concludes that a cooperative venture is desirable after reading the introductory material, the participant should contact the NIST Technology Transfer Office. He will then be referred to an appropriate laboratory scientist.

For more information about NIST, contact:

Main Office: (301) 975-3084
Technology Development and Small Business: (301) 975-3804
Grants and Funding: (301) 975-6328
Publications: (301) 975-2768

RESEARCH TRIANGLE INSTITUTE (RTI)

The RTI is designed to bring industry and government together to transfer and commercialize technology. It attributes its success to its proactive, interactive, and problem solving capabilities. RTI develops program outreach through brochures and campaigns promoting technology transfer programs and commercial opportunity workshops. It also identifies promising technologies by assessing industry's need, and analyzes their commercial potential. Last, the RTI advises on cooperative projects and maintains a database of technologies, patents, contacts, and spin-off benefit called TechTracs.

The RTI staff members are recruited for their experience in both the public and private sectors. They all have expertise in product development, negotiations, and technology in a broad range of industries.

For more information about RTI, contact:

Dr. Doris Rouse
Center for Technology Application
P.O. Box 12194
Research Triangle Park, NC 27709-2194
Phone: (919) 541-6980

TECHNOLOGY REINVESTMENT PROJECT (TRP)

TRP is a federal program that distributes funds to industry for research and development through grant competitions.

Its primary instrument for outreach is a series of workshops for both industry and government on cooperative ventures. Charts used in these workshops are available through the National Technology Information Service. The order form for these charts and a complete briefing on the TRP are in the basic information packet.

TRP markets its resources through news releases drafted by the Pentagon's Office of Public Affairs, articles in trade magazines, networking workshops, and solicitations through its free mailing list. However, TRP does not actively seek industry partners; they must actively pursue opportunity.

In order to receive information from TRP, interested parties must fax or mail a request or question which will be given to the appropriate person who will then respond within a day. TRP personnel will not provide initial information over the telephone.

For more information about the TRP contact:

Phone: (800) DUAL-USE
Fax: (703) 696-3813

ADVANCED TECHNOLOGY PROGRAM (ATP)

The ATP funds technology developed by industry. To distribute these funds, the ATP holds competitions for funding proposals. The ATP does not, however, advise on cooperative ventures, so they do not have information regarding CRDAs or possible government partners.

ATP advertises in the *Commerce Business Daily*, participates in workshops about technology transfer, distributes announcements through their mailing list, and lists its available funds in the *Federal Register*.

For more information about the ATP contact:

Phone: (301) 975-2636
Hot-line: (800) ATP-FUND
Recorded message: (301) 975-2273

GREAT LAKES INDUSTRIAL TECHNOLOGY CENTER (GLITeC)

GLITeC is the NASA Midwest Regional Technology Transfer Center. As one of six regional centers, GLITeC makes NASA technology and expertise available to industry in the Great Lakes region. GLITeC offers technology problem solving and commercialization services, including needs assessments, identification and referrals to technical expertise, and technology evaluations.

For more information, contact GLITeC at:
Phone: (216) 734-0094

WRIGHT TECHNOLOGY NETWORK (WTN)

Located in Dayton, Ohio, the Wright Technology Network (WTN) is a not-for-profit corporation established to facilitate the effective transfer and profitable commercialization of technology from federal laboratories and academic institutions to industries in the Great Lakes region. WTN collaborates with other organizations nationally and retains special emphasis on the technologies available at Wright Patterson Air Force Base (WPAFB).

WTN focuses on technology transfer from a "technology- push" and "technology-pull" perspective. It supports the transfer of technology through a variety of mechanisms, including Cooperative Research and Development Agreements, Technical Assistance Projects, Commercialization Projects, and Partnerships.

For more information, contact the WTN at:
Phone: (937) 253-0217
Toll-Free: (800) 240-8324
Fax: (937) 253-7238
www.wtn.org

CENTER FOR AEROSPACE INFORMATION (CASI)

CASI, a division of NASA, sponsors the Science and Technical Information (STI) program which offers a database of more than three million citations from journals and aerospace-related publications. This database allows direct, on-line access and automatic document distribution. The CASI brochure includes information on products, services, STI national contact people, and the STI registration form.

For more information about CASI, contact:
NASA Access Help Desk: (301) 621-0390
Director of STI Program: (703) 271-5640
Commercialization Program: (301) 621-0241

For more information about NASA, contact:
NASA Center for AeroSpace Information
Attn: Registration Services
800 Elkridge Landing Road
Lincoln Heights, MD 21090-2934
Phone: (301) 621-0153.

SMALL BUSINESS INNOVATION RESEARCH (SBIR) PROGRAM

The SBIR Program funds technology developed by small businesses. Each of the 11 largest government agencies which participate in cooperative R&D are mandated by law to set aside a minimum of 1.25% of their extramural R&D budgets to be awarded to small business proposals for innovation.

The SBIR programs are conducted in three phases. Phase I awards up to $100,000 for periods of up to six months to projects considered to have scientific technical merit and feasibility. Phase II provides up to $750,000 for a period of up to two years to winners of Phase I so they may continue their project development. Phase III commercializes the results of

Phase II. SBIR programs will not provide direct funding for this phase but will assist projects with finding alternative sources.

The SBIR programs in each of the agencies are conducted and reviewed by the Small Business Administration (SBA). It also maintains a mailing list of small businesses and publishes pre-solicitation announcements. More importantly, the SBA runs the Commercialization Matching System (CMS) which links potential sources of private-sector venture capital to SBIR awardees in order to fund Phase III commercialization.

For more information about the SBIR programs, contact:

Small Business Administration
Office of Technology
Phone: (202) 205-6450
Fax: (202) 205-7754

To be placed on the pre-solicitation mailing list, contact:

Automated phone: (202) 205-7777

TECHNOLOGY TRANSFER SOCIETY

The Technology Transfer Society, under the office of Bruce Becker, Executive Director, was founded in 1975 to foster the understanding and exchange of ideas on the best ways for new technology to reach industry, consumers and commercial markets. Professional society publishes a quarterly peer-reviewed journal and monthly newsletters. For more information, contact

(312) 644-0828 ext. 216

BRIDGING OFFICES OF VARIOUS DEPARTMENTS AND AGENCIES OF THE GOVERNMENT

Department of Defense

The office of Dr. Lance A. Davis, Defense Research and Engineering, monitors all DoD R&D activities to identify technologies that have potential non-defense commercialization application, assists private firms in resolving policy issues involved with technology transfer. For more information, contact

Cynthia E. Gonsalves
gonsalce@acq.osd.mil
(703) 681-5459

Department of Commerce

The office of Mary L. Good, Under Secretary for Technology, oversees several programs which include the Manufacturing Extension Partnership (800-MEP-4MFG), the Advanced Technology Program (800-ATP-FUND), the Partnership for a New Generation of Vehicles initiative (202-482-6268 or dnewell@doc.gov), and the National Institute of Standards and Technology (301-975-3084 or ipp@nist.gov). For more information, visit the web page at

www.ta.doc.gov

Department of Transportation

The office of Dr. Fenton Carey, Associate Administrator, coordinates DOT laboratories currently working in partnership opportunities in transportation information infrastructure, next-generation vehicles, and physical infrastructure. For more information, contact the DOT at

Phone: (202) 366-4978
t2.dot.gov or www.tsp.dot.gov

Department of Agriculture

The office of Dr. Richard M Parry, Jr., Assistant Administrator, Office of Technology Transfer, manages the USDA's Agriculture Research Service seeks CRADAs, patent, and licensing opportunities to lever commercial productivity in agriculture businesses. For more information, contact the USDA at

Phone: (301) 504-5345,
Fax: (301) 504-5060

Environmental Protection Agency

The office of Larry Fradkin, Director, Federal Tech Transfer Act Program, is interested in cost-effective technologies to prevent and control pollution. For more information, contact

Phone: (513) 569-7960
Fax: (513) 569-7132
fradkin.larry@epamail.epa.gov

US Geological Survey

The office of Gordon P. Eaton, Director, directs the USGS focus on eight business activities: water availability and quality, natural hazards, geographic and cartographic information, contaminated environments, land and water use, nonrenewable resources, environmental effects, and biological resources. For more information, contact

Phone: (703) 648-4450
www.usgs.gov

Department of Interior, Bureau of Reclamation

The office of Eluid Martinez, Commissioner, directs the R&D program supporting the Bureau's mission of managing, developing, and protecting water and related resources with emphasis on water augmentation, materials engineering, applied science, hydraulics, and electric power. For more information, contact

Phone: (202) 208-5671

Small Business Administration

The office of Aida Alvarez, who heads the SBA's Office of Technology, has three initiatives linked to high technology including: Small Business Innovation Research (SBIR), Small Business Technology Transfer Pilot Program (STTR), and the Research and Development Goaling Program. For more information, contact

Phone: (202) 205-6740
Fax: (202) 205-6901

National Association of Counties

Michael Hightower, Commissioner, Fulton County Georgia, is president of NACo. The objectives of NACo include: providing resources to help counties find innovative methods to meet technological and other challenges, liaison to other levels of government, and serve as national advocate for counties. For more information, contact

Phone: (202) 942-4248
www.naco.org

Appendix G

Model CRDA

The following pages are excerpted from Air Force Instruction 61-302 on COOPERATIVE RESEARCH AND DEVELOPMENT AGREEMENTS dated 8 June 1994. The first part gives the Guidelines for Using the AF Model CRDA. The next part is the running text of the Model Cooperative Research and Development Agreement and its Appendix A: Work Statement. This appendix concludes with a checklist designed to help the negotiating authority ensure that all relevant issues have been addressed.

GUIDELINES FOR USING THE AF MODEL CRDA

A1.1. Purpose of the Model. The Air Force has formulated the Model CRDA to minimize the amount of composition a preparer must do to construct an effective CRDA. The Model CRDA provides a recommended format and specific terms and conditions consistent with law, regulation, and policy.

A1.2. Guidelines for Preparation.

A1.2.1. CRDA Number. The CRDA number contains a two-digit fiscal year designation, a two-letter Air Force Laboratory designation, and a two-digit serial number (for example, 93-WL-01). The office of primary responsibility (OPR) for CRDAs, which is the local Office of Research and Technology Application (ORTA) at each installation, assigns the number.

A1.2.2. Caption and Title. Enter the name of the Air Force activity and the corporate name(s) of the collaborator(s).

A1.2.3. Article 1: Preamble. Enter:

— The name and address of the collaborating party.

— The name, functional address symbol, and address of the participating Air Force organization.

A1.2.3.1. The model CRDA refers to the collaborating party as the Collaborator and to the participating Air Force organization as the Air Force Activity. Use these terms in the CRDA. In CRDAs with more than two collaborators, modify the articles accordingly. You may also substitute the actual names of the Air Force Activity and the Collaborator.

A1.2.4. Article 2: Definitions. The definitions article establishes the meanings of specified terms to be consistent with Federal laws and regulations. Use the definitions exactly as they appear in the model. When necessary, you may include definitions for additional terms.

A1.2.5. Article 3: Work Statement. This describes the nature and scope of the work performed under this agreement. It details requirements for equipment, maintenance and other support, and reporting.

A1.2.6. Article 4: Financial Obligations. State the Collaborator's funding obligations and identify the Air Force Activity's financial office and address to which the collaborator should send payments.

A1.2.6.1. For payments resulting from Article 4, paragraphs 4.1 or 4.2, enter the Air Force activity's organization symbol and financial office address (for example, ASC/AFO, Wright-Patterson AFB, OH 45433-5000).

A1.2.6.2. Comply with the patents article in granting royalty or other income from the licensing of patents.

A1.2.7. **Article 5: Patents. This article sets out the disposition of rights with respect to ownership, filing of patent applications, and granting of licenses in inventions made under the CRDA. For any proposed CRDA that could result in an invention, include this article exactly as written. AFLSA/JACP, JACPD or JACPB prepares any patent license agreement under Article 5, paragraph 5.2. (See AFI 61-303.)

A1.2.8. **Article 6: Copyrights. This article sets out the disposition of rights with respect to ownership and granting of licenses in copyrightable works created under the CRDA. The royalty percentages set out in this article are suggested rates; you may change them on a case-by-case basis. Otherwise, include this article exactly as written.

A1.2.9. **Article 7: Proprietary Information. If the proposed CRDA involves proprietary information, include provisions for its protection (15 U.S.C. 3710a[c][7]). Include this article exactly as written. See Article 2 for a definition of proprietary information.

***NOTE:** Any CRDA that includes rights in any background intellectual property (that is, patents, copyrights, and proprietary information) may require additional clauses in the CRDA, separate license agreements, or both. Coordinate any changes to the patents, copyrights, or proprietary information articles with Air Force Patent Counsel.

A1.2.10. Article 8: Term, Modification, Extension, Termination, and Disputes. Enter the term of the proposed CRDA in the first paragraph of this article. Otherwise, copy the article exactly as written.

A1.2.11. Article 9: Representations and Warranties. This article sets out certain representations and warranties by the parties. Tailor the representations and warranties of the collaborator as appropriate in consideration of the identity of the collaborator.

A1.2.12. Article 10: Liability. This article sets out the liability of the Government for injury to persons or property in the course of work under the CRDA. Prominently display the "No Warranty" clause (using bold type or all caps, for example) to meet Uniform Commercial Code requirements that you duly notify parties of any disclaimers. Include this article exactly as written.

A1.2.13. Article 11: General Terms and Provisions. This Article sets out various miscellaneous contract provisions required for the proposed CRDA. Include this article exactly as written.
A1.2.14. Article 12: Notices. Enter the names and addresses of the persons who are to receive notices under the CRDA related to formal contract matters and technical matters.

A1.2.15. Signatures. The Air Force activity commander, director, or other authorized person and an authorized representative of the collaborator sign the CRDA. (The Air Force reviewing official may approve, disapprove, or require modification to the CRDA within 30 days of receipt of a signed agreement.)

A1.2.16. Appendix A: Work Statement. This lists the requirements and suggested format.

MODEL COOPERATIVE RESEARCH AND DEVELOPMENT AGREEMENT

USAF CRDA NUMBER___________

USAF COOPERATIVE RESEARCH AND DEVELOPMENT AGREEMENT

BETWEEN

[Insert name of Air Force Activity]

and

[Insert name of collaborator]

Article 1. Preamble

1.1 This Cooperative Research and Development Agreement (*Agreement*) for performing the work described in the Work Statement attached hereto as Appendix A is entered into pursuant to 15 U.S.C. § 3710a (as amended) and Air Force Policy Directive 61-3 by and between **[Insert name of collaborator]**, (hereinafter referred to as "*Collaborator*"), located at **[Insert address of collaborator]**, and the United States of America as represented by the Department of the Air Force, (**[Insert name of Air Force Activity]**), (hereinafter referred to as the "*Air Force Activity*"), located at ___ Air Force Base, (State). The terms and conditions of this *Agreement* are set forth as follows.

Article 2. Definitions

2.1 As used in this *Agreement*, the following terms shall have the following meanings and such meanings shall be applicable to both the singular and plural forms of the terms:

2.2 "*Created*" in relation to any copyrightable work means when the work is fixed in any tangible medium of expression for the first time, as provided for at 17 U.S.C. § 101.

2.3 "*Effective Date*" means the earlier of: (a) the date of the last signature of the duly authorized representatives of the parties and the *Reviewing Official*; or (b) thirty (30) days after the receipt of a signed copy of this *Agreement* by the *Reviewing Official* without that official taking any action thereon.

2.4 "*Government*" means the Government of the United States of America.

2.5 "*Government Purpose License*" or "*GPL*" means a license to the *Government* conveying a nonexclusive, irrevocable, worldwide, royalty-free license to practice and have practiced an *Invention* for or on behalf of the *Government* for government purposes and on behalf of any foreign government or international organization pursuant to any existing or future treaty or agreement with the United States, and conveying the right to use, duplicate or disclose copyrighted works or *Proprietary Information* in whole or in part and in any manner, and to have or permit others to do so, for Government purposes. Government purposes include competitive procurement, but do not include the right to have or permit others to practice an *Invention* or use, duplicate or disclose copyrighted works or *Proprietary Information* for commercial purposes.

2.6 "*Invention*" means any invention or discovery that is or may be patentable or otherwise protected under Title 35 of the United States Code or any novel variety of plant which is or may be protectable under the Plant Variety Protection Act (7 U.S.C. § 7321 et seq.).

2.7 "*Made*" in relation to any *Invention* means the conception or first actual reduction to practice of such *Invention*.

2.8 "*Proprietary Information*" means information which embodies trade secrets or which is confidential technical, business or financial information provided that such information:

i) is not generally known, or is not available from other sources without obligations concerning its confidentiality;

ii) has not been made available by the owners to others without obligation concerning its confidentiality;

iii) is not described in an issued patent or a published copyrighted work or is not otherwise available to the public without obligation concerning its confidentiality; or

iv) can be withheld from disclosure under 15 U.S.C. § 3710a(c)(7)(A) & (B) and the Freedom of Information Act, 5 U.S.C. § 552 et seq.; and

v) is identified as such by labels or markings designating the information as proprietary.

2.9 "*Reviewing Official*" means the authorized representative of the Department of the Air Force who is identified on the signature page of this *Agreement*.

2.10 "*Under*" as used in the phrase "*Under this Agreement*" means within the scope of work performed under this *Agreement*.

Article 3. Work Statement

3.1 Appendix A sets forth the nature and scope of the work performed *Under this Agreement*, including any equipment, maintenance and other support, and any associated reporting requirements.

3.2 The *Collaborator* may inspect *Government* property identified in Appendix A prior to use. Such property may be repaired or modified at the *Collaborator's* expense only after obtaining the written approval of the *Air Force Activity*. Any repair or modification of the property shall not affect the title of the *Government*. Unless *Air Force Activity* hereafter otherwise agrees, the *Collaborator* shall, at no expense to the *Air Force Activity*, return all *Government* property after termination or expiration of this *Agreement* in the condition in which it was received, normal wear and tear excepted.

3.3 The parties agree to confer and consult with each other prior to publication or other public disclosure of the results of work *Under this Agreement* to ensure that no *Proprietary Information* or military critical technology or other controlled information is released. Prior to submitting a manuscript for publication or before any other public disclosure, each party will offer the other party ample opportunity to review such proposed publication or disclosure, to submit objections, and to file applications for patents in a timely manner.

Article 4. Financial Obligations

4.1 The *Collaborator* will pay the *Air Force Activity* the amount of $within thirty (30) days after the *Effective Date* hereof. Subsequent payments will be paid as follows:.

4.2 Payments from copyrights shall be payable by the *Collaborator* to the *Air Force Activity* in accordance with the provisions of Article 6.

4.3 Except as provided for in paragraph 4.4, payments by the *Collaborator* to the *Air Force Activity* under this Article shall be made payable to the *Air Force Activity* and mailed to the following address:
[Insert appropriate *Air Force Activity's* accounting office FAS], Air Force Base, (State, ZIP)

4.4 Royalty or other income from patents shall be payable in accordance with any patent license under Article 5.

Article 5. Patents

5.1 **Disclosure of *Inventions***. Each party shall report to the other party, in writing, each *Invention Made Under this Agreement*, promptly after the existence of each such *Invention*, in the exercise of reasonable diligence, becomes known.

5.2 **Rights in *Inventions***. Each party shall separately own any *Invention Made* solely by its respective employees *Under this Agreement*. *Inventions Made* jointly by the *Air Force Activity* and the *Collaborator* employees shall be jointly owned by both parties. The *Collaborator* shall have an option under 15 U.S.C. 3710a(b)(2) to obtain an exclusive or non-exclusive license at a reasonable royalty rate, subject to the retention of a *GPL* by the *Government*, in any *Invention Made* by the *Air Force Activity* employees *Under this Agreement*. The *Collaborator* shall exercise the option to obtain a license by giving written notice thereof to the *Air Force Activity* within three (3) months after disclosure of the *Invention* under paragraph 5.1. The royalty rate and other terms and conditions of the license shall be set forth in a separate license agreement and shall be negotiated promptly after notice is given. The *Collaborator* hereby grants to the *Government*, in advance, a *GPL* in any *Invention Made* by the *Collaborator* employees *Under this Agreement*.

5.3 **Filing Patent Applications**. The *Collaborator* shall have the first option to file a patent application on any *Invention Made Under this Agreement*, which option shall be exercised by giving notice in writing to the *Air Force Activity* within three (3) months after disclosure of the *Invention* under paragraph 5.1, and by filing a patent application in the U.S. Patent and Trademark Office within six (6) months after written notice is given. If the *Collaborator* elects not to file or not to continue prosecution of a patent application on any such *Invention* in any country or countries, the *Collaborator* shall notify the *Air Force Activity* thereof at least three (3) months prior to the expiration of any applicable filing or response deadline, priority period or statutory bar date. In any country in which the *Collaborator* does not file, or does not continue prosecution of, or make any required payment on, an application on any such *Invention*, the *Air Force Activity* may file, or continue prosecution of, or make any required payment on, an application, and the *Collaborator* agrees, upon request by the *Air Force Activity*, to assign to the *Government* all right, title and interest of the *Collaborator* in any such application and to cooperate with the *Air Force Activity* in executing all necessary documents and obtaining cooperation of its employees in executing such documents related to such application. The party filing an application shall provide a copy thereof to the other party. ***NOTE:*** Any patent application filed on any *Invention Made Under this Agreement* shall include in the patent specification thereof the statement: "This invention was made in the performance of a cooperative research and development agreement with the Department of the Air Force. The invention may be manufactured and used by or for the Government of the United States for all government purposes without the payment of any royalty."

5.4 **Patent Expenses**. Unless otherwise agreed, the party filing an application shall pay all patent application preparation and filing expenses and issuance, post issuance and patent maintenance fees associated with that application.

Article 6. Copyrights

6.1 The *Collaborator* shall own the copyright in all works *Created* in whole or in part by the *Collaborator Under this Agreement*, which are copyrightable under Title 17, United States Code. The *Collaborator* shall mark any such works with a copyright notice showing the *Collaborator* as an owner and shall have the option to register the copyright at the *Collaborator's* expense.

6.2 The *Collaborator* hereby grants in advance to the *Government* a GPL in all copyrighted works *Created Under this Agreement*. The *Collaborator* will prominently mark each such copyrighted work subject to the *GPL* with the words: "This work was created in the performance of a Cooperative Research and Development Agreement with the Department of the Air Force. The Government of the United States has a royalty-free Government purpose license to use, duplicate or disclose the work, in whole or in part and in any manner, and to have or permit others to do so, for Government purposes."

6.3 The *Collaborator* shall furnish to the *Air Force Activity*, at no cost to the *Air Force Activity*, three (3) copies of each work *Created* in whole or in part by the *Collaborator Under this Agreement*.

6.4 The *Collaborator* shall pay to the *Air Force Activity* twenty percent (20%) of all gross income received by the *Collaborator* or its affiliates from the sale, lease or rental of any copyrighted work *Created Under this Agreement*. The *Collaborator* shall pay to the *Air Force Activity* fifty percent (50%) of all gross income or royalties received by the *Collaborator* or its affiliates from the licensing or assignment of any copyrighted work *Created Under this Agreement*. Any sale, lease or rental to *Government* shall not be subject to payments hereunder and shall be discounted in price by a corresponding amount. All such payments to the *Air Force Activity* shall be due and paid on or before the last day of the month next following receipt by the *Collaborator* of any such gross income or gross royalties. The *Collaborator* shall provide to the *Air Force Activity* a report at least annually showing all gross income and royalties received. The *Collaborator* shall make payments due hereunder to the *Air Force Activity* in accordance with paragraph 4.3 of this *Agreement*. The *Collaborator's* obligation to make payments to the *Air Force Activity* hereunder shall survive expiration or other termination of this *Agreement*.

6.5 The *Air Force Activity*, at its expense, may require an accounting of income received by the *Collaborator* and its affiliates and may, at reasonable times and upon reasonable notice to the *Collaborator*, examine the *Collaborator*'s and any affiliate's books and records to verify the accounting.

Article 7. Proprietary Information

7.1 Neither party to this *Agreement* shall deliver to the other party any *Proprietary Information* not developed *Under this Agreement*, except with the written consent of the receiving party. Unless otherwise expressly provided in a separate document, such *Proprietary Information* shall not be disclosed by the receiving party except under a written agreement of confidentiality to employees and contractors of the receiving party who have a need for the information in connection with their duties *Under this Agreement*.

7.2 *Proprietary Information* developed *Under this Agreement* shall be owned by the developing party, and any jointly developed *Proprietary Information* shall be jointly owned. *Government* shall have a GPL to use, duplicate and disclose, in confidence, and to authorize others to use, duplicate and disclose, in confidence, for government purposes, any such *Proprietary Information* developed solely by the *Collaborator*. The *Collaborator* may use, duplicate and disclose, in confidence, and authorize others on its behalf to use, duplicate and disclose, in confidence, any such *Proprietary Information* developed solely by the *Air Force Activity*. *Proprietary Information* developed *Under this Agreement* shall be exempt from the Freedom of Information Act, 5 U.S.C. § 552 et seq., as provided at 15 U.S.C. § 3710a(c)(7)(A) & (B). The exemption for *Proprietary Information* developed jointly by the parties or solely by the *Air Force Activity* shall expire not later than five years from the date of development of such *Proprietary Information*.

Article 8. Term, Modification, Extension, Termination and Disputes

8.1 **Term and Extension**. The term of this *Agreement* is for a period of ( ) months, commencing on the *Effective Date* of this *Agreement*. This *Agreement* shall expire at the end of this term unless both parties hereto agree in writing to extend it further. Expiration of this *Agreement* shall not affect the rights and obligations of the parties accrued prior to expiration.

8.2 **Modification**. Any modifications shall be by mutual written agreement signed by the parties' representatives authorized to execute this *Agreement* and attached hereto. A copy of any modifications will be forwarded to the *Reviewing Official* for information purposes.

8.3 **Termination**. Either party may terminate this *Agreement* for any reason upon delivery of written notice to the other party at least three (3) months prior to such termination. Termination of this *Agreement* shall not affect the rights and obligations of the parties accrued prior to the date of termination of this *Agreement*. In the event of termination by either party, each party shall be responsible for its own costs incurred through the date of termination, as well as its own costs incurred after the date of termination and which are related to the termination. If the *Air Force Activity* terminates this *Agreement*, it shall not be liable to the *Collaborator* or its contractors or subcontractors for any costs resulting from or related to the termination, including, but not limited to, consequential damages or any other costs.

8.4 **Disputes**. All disputes arising out of, or related to, this *Agreement* shall be resolved in accordance with this Article.

8.4.1 The parties shall attempt to resolve disputes between themselves. Any dispute which is not disposed of by agreement of the parties shall be referred to the *Reviewing Official* for decision.

8.4.2 *Reviewing Official*. The *Reviewing Official* shall within sixty (60) days of the receipt of the dispute, notify the parties of the decision. This decision shall be final and conclusive unless, within thirty (30) days from the date of receipt of such copy, either party submits to the *Reviewing Official*, a written appeal addressed to the Secretary of the Air Force.

8.4.3 Secretary of the Air Force. The decision of the Secretary of the Air Force, or the Secretary's duly authorized representative, on the appeal shall be final and conclusive.

8.5 Continuation of Work. Pending the resolution of any such dispute, work under this *Agreement* will continue as elsewhere provided herein.

Article 9. Representations and Warranties

9.1 The *Air Force Activity* hereby represents and warrants to the *Collaborator* as follows:

9.1.1 **Mission**. The performance of the activities specified by this *Agreement* are consistent with the mission of the *Air Force Activity*.

9.1.2 **Authority**. All prior reviews and approvals required by regulations or law have been obtained by the *Air Force Activity* prior to the execution of the *Agreement*. The *Air Force Activity* official executing this *Agreement* has the requisite authority to do so.

9.1.3 **Statutory Compliance**. The *Air Force Activity*, prior to entering into this *Agreement*, has (1) given special consideration to entering into cooperative research and development agreements with small business firms and consortia involving small business firms; (2) given preference to business units located in the United States which agree that products embodying an *Invention Made* under this *Agreement* or produced through the use of such *Invention* will be manufactured substantially in the United States; and (3) taken into consideration, in the event this *Agreement* is made with an industrial organization or other person subject to the control of a foreign company or government, whether or not such foreign government permits United States agencies, organizations, or other persons to enter into cooperative research and development agreements and licensing agreements with such foreign country.

9.2 The *Collaborator* hereby represents and warrants to the *Air Force Activity* as follows:

9.2.1 **Corporate Organization**. The *Collaborator*, as of the date hereof, is a corporation duly organized, validly existing and in good standing under the laws of the State of [insert state], and (if applicable) is a wholly owned subsidiary of [insert parent corp name], a [insert state] corporation.

9.2.2 **Statement of Ownership**. The *Collaborator* (is) (is not) a foreign owned or a subsidiary of a foreign-owned entity. The *Collaborator* has the right to assignment of all *Inventions Made* and copyrightable works *Created* by its employees *Under this Agreement*.

9.2.3 **Authority**. The *Collaborator* official executing this *Agreement* has the requisite authority to enter into this *Agreement* and the *Collaborator* is authorized to perform according to the terms thereof.

Article 10. Liability

10.1 **Property**. All property is to be furnished "as is." No party to this *Agreement* shall be liable to any other party for any property of that other party consumed, damaged or destroyed in the performance of this *Agreement*, unless it is due to the gross negligence or willful misconduct of the party or an employee or agent of the party.

10.2 ***Collaborator* Employees**. The *Collaborator* agrees to indemnify and hold harmless and defend the *Government*, its employees and agents, against any liability or loss for any claim made by an employee or agent of the *Collaborator*, or persons claiming through them, for death, injury, loss or damage to their person or property arising in connection with this *Agreement*, except to the extent that such death, injury, loss or damage arises solely from the negligence of the *Air Force Activity* or its employees.

10.3 **NO WARRANTY**. EXCEPT AS SPECIFICALLY STATED IN ARTICLE 9, OR IN A LATER *AGREEMENT*, THE PARTIES MAKE NO EXPRESS OR IMPLIED WARRANTY AS TO ANY MATTER WHATSOEVER, INCLUDING THE CONDITIONS OF THE RESEARCH OR ANY INVENTION OR PRODUCT, WHETHER TANGIBLE OR INTANGIBLE, MADE, OR DEVELOPED *UNDER THIS AGREEMENT*, OR THE MERCHANTABILITY, OR FITNESS FOR A PARTICULAR PURPOSE OF THE RESEARCH OR ANY INVENTION OR PRODUCT. THE PARTIES FURTHER MAKE NO WARRANTY THAT THE USE OF ANY INVENTION OR OTHER INTELLECTUAL PROPERTY OR PRODUCT CONTRIBUTED, MADE OR DEVELOPED *UNDER THIS AGREEMENT* WILL NOT INFRINGE ANY OTHER UNITED STATES OR FOREIGN PATENT OR OTHER INTELLECTUAL PROPERTY RIGHT. IN NO EVENT WILL ANY PARTY BE LIABLE TO ANY OTHER PARTY FOR COMPENSATORY, PUNITIVE, EXEMPLARY, OR CONSEQUENTIAL DAMAGES.

10.4 **Other Liability**. The *Government* shall not be liable to any other party to this *Agreement*, whether directly or by way of contribution or indemnity, for any claim made by any person or other entity for personal injury or death, or for property damage or loss, arising in any way from this *Agreement*, including, but not limited to, the later use, sale or other disposition of research and technical developments, whether by resulting products or otherwise, whether made or developed *Under this Agreement*, or whether contributed by either party pursuant to this *Agreement*, except as provided under the Federal Tort Claims Act (28 U.S.C. §§ 2671 et seq.) or other Federal law where sovereign immunity has been waived.

Article 11. General Terms and Provisions

11.1 **Disposal of Toxic or Other Waste**. The *Collaborator* shall be responsible for the removal and disposal from the *Air Force Activity* property of any and all toxic or other material provided or generated by the *Collaborator* in the course of performing this *Agreement*. The *Collaborator* shall obtain at its own expense all necessary permits and licenses as required by local, state, and Federal law and regulation and shall conduct such removal and disposal in a lawful and environmentally responsible manner.

11.2 **Force Majeure**. Neither party shall be in breach of this *Agreement* for any failure of performance caused by any event beyond its reasonable control and not caused by the fault or negligence of that party. In the event such a force majeure event occurs, the party unable to perform shall promptly notify the other party and shall in good faith maintain such part performance as is reasonably possible and shall resume full performance as soon as is reasonably possible.

11.3 **Relationship of the Parties**. The parties to this *Agreement* and their employees are independent contractors and are not agents of each other, joint venturers, partners or joint parties to a formal business organization of any kind. Neither party is authorized or empowered to act on behalf of the other with regard to any contract, warranty or representation as to any matter, and neither party will be bound by the acts or conduct of the other. Each party will maintain sole and exclusive control over its own personnel and operations.

11.4 **Publicity/Use of Name Endorsement**. Any public announcement of this *Agreement* shall be coordinated between the *Collaborator*, the *Air Force Activity* and the public affairs office supporting the *Air Force Activity*. The *Collaborator* shall not use the name of the *Air Force Activity* or the *Government* on any product or service which is directly or indirectly related to either this *Agreement* or any patent license or assignment which implements this *Agreement* without the prior written approval of the *Air Force Activity*. By entering into this *Agreement*, the *Air Force Activity* or the *Government* does not directly or indirectly endorse any product or service provided, or to be provided, by *Collaborator*, its successors, assignees, or licensees. The *Collaborator* shall not in any way imply that this *Agreement* is an endorsement of any such product or service.

11.5 **No Benefits**. No member of, or delegate to the United States Congress, or resident commissioner, shall be admitted to any share or part of this *Agreement*, nor to any benefit that may arise therefrom; but this provision shall not be construed to extend to this *Agreement* if made with a corporation for its general benefit.

11.6 **Governing Law**. The construction, validity, performance and effect of this *Agreement* for all purposes shall be governed by the laws applicable to the *Government*.

11.7 **Waiver of Rights**. Any waiver shall be in writing and provided to all other parties. Failure to insist upon strict performance of any of the terms and conditions hereof, or failure or delay to exercise any rights provided herein or by law, shall not be deemed a waiver of any rights of any party hereto.

11.8 **Severability**. The illegality or invalidity of any provisions of this *Agreement* shall not impair, affect or invalidate the other provisions of this *Agreement*.

11.9 **Assignment**. Neither this *Agreement* nor any rights or obligations of any party hereunder shall be assigned or otherwise transferred by any party without the prior written consent of all other parties.

11.10 **Controlled Information**. The parties understand that information and materials provided pursuant to or resulting from this *Agreement* may be export controlled, classified, or unclassified sensitive and protected by law, executive order or regulation. Nothing in this *Agreement* shall be construed to permit any disclosure in violation of those restrictions.

Article 12. Notices

12.1. Notices, communications, and payments specified in this agreement shall be deemed made if given and addressed as set forth below.

A. Send formal notices under this agreement by prepaid certified U.S. Mail and address them:

Air Force Activity:	Attn: (ORTA)
Address	
Collaborator:	Attn:
Address	

B. Send correspondence on technical matters by prepaid ordinary U.S. Mail and address them:

Air Force Activity:	Attn:
Address	
Collaborator:	Attn:
Address	

IN WITNESS WHEREOF, the Parties have executed this agreement in duplicate through their duly authorized representatives as follows:

COLLABORATING PARTY	**AIR FORCE ACTIVITY**
(Name of Collaborating Activity)	(Name of Air Force Activity)
(Name of Official—Printed or Typed)	(Name of Official—Printed or Typed)
(Signature of Official)	(Signature of Official)
(Title of Official)	(Title of Official)
(Address of Official)	(Address of Official)
(Date Signed)	(Date Signed)

REVIEWED AND APPROVED BY AIR FORCE REVIEWING OFFICIAL:

Name of Air Force Reviewing Official—Printed or Typed)

(Title—Printed or Typed)

(Signature) (Date)

APPENDIX A: WORK STATEMENT

NOTE: This is an essential part of the CRDA and should be completed first. It should:
—Describe the scope of work in technical terms.
—Specify each party's research and development responsibilities.
—Detail each party's contribution of funds, personnel, services, property, equipment, and facilities.
—State each party's division of responsibilities for reporting progress and results.
—Identify the principal investigators for each party, the milestones for work progress, and the procedures for interaction between parties.
—Identify each party's background intellectual property rights and environmental, health, and safety responsibilities. This is particularly important if the CRDA may call for an exchange of materials, equipment, or facility use.

Include proprietary information (if any) in a separate, appropriately marked document.

Suggested Format:

1.0 Title. Provide a descriptive title of the CRDA.

2.0 Objective: State the overall purpose of the CRDA, including a short description of benefits anticipated for the Government and the collaborating party. Indicate the the thrust of the outcome, whether it is a product, process, facility, or personnel. 3.0 Background: Include any pertinent historical information related to the proposed CRDA. State each party's background intellectual property rights in either party and each party's responsibilities regarding health, safety, and environmental protection.

4.0 Technical Tasks: This section may include the following parts.

4.1 *Collaborator*: Describe the tasks that the *Collaborator* is to do and describe and estimate the value of the resources it is to provide in the form of funds, personnel, services, property, and equipment.

4.2 *Air Force Activity*: Describe the tasks that the *Air Force Activity* is to do and describe and estimate the value of the resources it is to provide in the form of personnel, services, property, and equipment.

5.0 Deliverables or Desired Benefits: This section may include the following.

5.1 Benefits to the Collaborating Party: Describe what the collaborating party hopes to accomplish and how the collaborating party plans to benefit, directly or indirectly, from the CRDA.

5.2 Benefits to the Government: Describe how the Government will benefit directly or indirectly from the CRDA.

6.0 Other: Give any other pertinent information that would help both parties understand their respective roles in the CRDA.

7.0 Milestones: Give the dates each party is expected to complete its tasks.

8.0 Reports: List the reports each party is to generate and give a schedule for their completion. Parties should prepare and submit written progress reports at least every six (6) months, and a final report within two (2) months after the CRDA or work under the CRDA ends. Parties should coordinate the formats.

CHECKLIST BASED ON THE MODULAR AIR FORCE CRDA

This section is designed to help the negotiating authority be sure that all relevant issues in negotiating the CRDA have been addressed.

CRDA Number. The local Office of Research and Technology Application (ORTA) will assign this number. The number format, in order, is a two-digit fiscal year designation, a two-letter Air Force Laboratory designation, and a two-digit serial number.

Caption and Title. Enter the name of the Air Force activity and the corporate name(s) of the collaborator(s).

Preamble. Enter the name and address of the collaborating party name, functional address symbol, and the address of the participating Air Force organization.

Definitions. This article establishes the meanings of specific terms to be consistent with federal laws and regulations. The existing definitions should be used exactly as they appear in the model. However, you may add additional definitions to cover other terms.

Work Statement, Responsibility, and Reporting Requirements. This section describes the nature and scope of the work to be performed under the agreement and each party's responsibilities. Because it is one of the most important parts of the agreement, it should be done first.

Financial Obligations. Enter the collaborator's funding obligations in this section, and identify the Air Force activity's financial office and address to which the collaborator should send payments.

Patents. In this section define the rights to any inventions that come out of the CRDA process.

Copyrights. This article sets out the disposition of rights with respect to ownership and permission to use works that can be copyrighted.

Proprietary Information. Include in this section any provisions for the protection of proprietary information.

Term, Modification, Extension, Termination, and Disputes. Enter the term of the proposed CRDA in the first paragraph of this section.

Representations and Warranties. This section carries certain representations and warranties by the parties.

Liability. This section establishes the liability of the Government for injury to persons or property in the course of work relating to the CRDA.

General Terms and Provisions. This article sets out miscellaneous contract provisions required for the proposed CRDA.

Notices. Enter the names and addresses of the persons who are to receive notices under the CRDA.

Signatures. The authorized persons representing the collaborator and the government agency must each sign. The CRDA will then be reviewed by a government official, who may approve, disapprove, or require modification of the CRDA before further consideration.

Appendix H

Technology Transfer Related Internet Sites

In Chapter 4: Building Bridges, we offer guidance to the technology transfer practitioner in the form of rules-of-thumb to apply when entering into cooperative R&D partnerships. Appendix F: Bridging Organizations, lists the various organizations and offices that oversee, manage, or serve as information brokerage houses for the technology transfer community. You can expect to find information on the web sites below that is much the same as that presented throughout this book. Because these web sites are "portals" to technology transfer information, you will likely find multiple links and connections between these sites. They represent, at this point in time, what we consider the "Best of the Web" list of technology transfer sites on the Internet.

BEST OF THE TECHNOLOGY TRANSFER BRIDGING SITES

Dozens of interesting sites fall under the "bridging organizations" category. Some information is easy to get to from organizational home pages. Other information requires a little "mining" within these sites. Those sites with the greatest "ease of use" and which get right at the technology transfer source information are as follows:

http://www.dtic.dla.mil/techtransit/techtransfer/com_t2.html

http://www.dtic.dla.mil/techtransit/techtransfer/bus_incu.html

http://www.dtic.dla.mil/natibo/ index.html http://www.dtic.dla.mil/techtransit/techtransfer/intemtl_t_2.html

http://www.dtic.dla.mil/techtransit/techtransfer/us_t_2.html

These first five sites should be grouped as one site because they are developed as part of and maintained by the Defense Technical Information Center (DTIC) home page. These sites have the most in-depth information and list in equal measure both the military and commercial organizations who are active in defense related technology transfer enterprises.

The New York Environmental & Energy Technology Exchange deserves a mention as one of the more user-friendly sites. This site offers ease of use and is a well-organized home page with information and links that makes mining for information easier to find and link to.

http://www.eba-nys.org/E2T2.html

Bio-diversity and Ecosystems Information site is well organized and is filled with links from international governments to local communities, along with links to other information services that are important to the technology exchange community.

http://straylight.tamu.edu/bene/bene.html

BEST OF THE TECH TRANSFER EDUCATION RELATED SITES

The Educator's TECH Exchange Magazine is the first site that needs mentioned here. It provides a magazine both in paper and Internet mediums. It is also available for free by filling out an online subscription form.

http://www.edtechx.com/

Georgia Tech Research Institute (GTRI) has one of the better sites available for links to their entire Internet directory. This site is well laid out and is one of the user-friendlier sites of the educational categories. After a visit here you will see why GTRI is ahead of its time in the educational field.

http://www.gtri.gatech.edu/ neurnet.html

Tech Transit is a site that is related to DTIC. The home page for this site allows you to choose from various sites that are available for you to search information on tech transfer. On the home page of this site you will be able to locate technology transfer resources for the Academic sites and other related material. DTIC sites in general are well maintained and easy to mine for the tech transfer information that you seek.

http://www.dtic.dla.mil/techtransit/ techtransfer/academic_t2.html

The Tech is a continuous news service that is offered by the Massachusetts Institute of Technology (MIT). This site has full browse and search capacities along with past and present archived material. The success of spin-off companies that have commercialized the intellectual property originating at MIT is legendary. If there is one school in the country that has "broken the code" on the technology exchange and commercialization process, it may be MIT.

http://www.the-tech.mit.edu/

BEST OF THE OTHER RELATED SITES

Once again the DTIC site comes out on top because of its numerous links to other technology transfer sites, databases, and information sources. Offices of Research and Technology Application (ORTAs), service directories, and links to other agencies are catalogued by DTIC.

http://www.dtic.dla.mil/techtransit/ techtransfer/orta_navg.html

http://www.dtic.dla.mil/techtransit/ websource/fedlabs_t2.html

http://www.dtic.dla.mil/techtransit/ techtransfer/techtrans.html

http://www.dtic.dla.mil/techtransit/ techtransfer/techtrans.html

http://www.dtic.dla.mil/techtransit/ techtransfer/poc.html

http://www.dtic.dla.mil/techtransit/ techtransfer/orta_state.html

http://www.dtic.dla.mil/labman/ projects/list.html

http://www.dtic.dla.mil/techtransit/ index.htm

http://www. dtic.dla.mil/techtransit/ bus_ops/partner_ops.html

Electronic Commerce/Electronic Data Interchange is sponsored by Key Software Solutions Inc. This site is filled with links to both public and private information along with a mail list that may be subscribed to via an online subscription request. The focus of this site is on international trade opportunities resultant from the North American Free Trade Agreement (NAFTA).

http://www.nafta.net/ecedi.html

Information Technology Resources Board (ITRB) maintains a comprehensive list of sites and contacts along with links to Government agencies. This site is easy to load on your Web browser and is geared to Information Technology (IT) articles.

http://www.itrb.fed.gov/

ACQweb is part of the office of the Under Secretary of Defense for Acquisitions and Technology. This site has all the components of navigating within the Acquisitions and Technology field. Technology transfer activities are subordinated to this DoD secretariat. This site, therefore, offers policy guidance, national trends, and important background information for those pursuing cooperative R&D partnerships.

http://www.acq.osd.mil/ HomePage.html

BEST OF THE INTERNATIONAL TECH TRANSFER RELATED SITES

Industry Link is one of the better-organized sites for searching information related to different industries and environmental concerns. This site also allows you to cross industries and connect to magnet sites. *Industry Link* is a first class site but slow to load if you have a slow connection speed.

http://www.industrylink.com/ alphaa-g.html

The Greentie Directory by far is one of the easiest sites to negotiate around. This site is filled with graphic pointers and text explaining what it is that you may be looking for and wish to research. The search engine and assisted search capability will find just about anything that you wish to find.

http://www.greentie.org/ grnfeed.html

Asia Pacific Foundation of Canada is an easy to use one page search interface that will allow the user to research the information by category of interest. The home page is graphical and loads to your web browser as fast as you can click its Internet address. This site is one of the easiest to use and negotiate.

http://www.apfnet.org/

Please note that all the web sites listed here in Appendix H were valid Internet sites as of September 1997. If you experience problems connecting to one of these site the problem may be that the particular server hosting a site may be down or that the site has recently been moved.

References

1. Lesko, John N. and Garvin, James T. "The R&D Coordinator: Functioning as a Technical Liaison Officer." Army, Research, Development, and Acquisition Bulletin. November–December 1989, pp. 21-23.

2. "What Companies Want From the Federal Labs," *Issues in Science and Technology*, Fall 1993.

3. "Not Invented Here: How US Industry Sources Technology from the Public Sector," Battelle Final Report to AFMC Battelle-ESI Team by Craig Chambers, May 1994.

4. *AFMC Technology Transfer Handbook*, Item F1.

5. Berger, Beverly. "Technology Transfer in a Time of Transition: A Guide to Defense Conversion." Federal Laboratory Consortium for Technology Transfer. January, 1994, Washington, DC.

6. McGregor, Douglas. "The Human Side of Enterprise." The Manager's Bookshelf: A Mosaic of Contemporary Views, Third Edition. Edited by Jon Pierce and John Newstrom. Harper Collins Publishers, 1993.

7. *Proceedings of Roundtable One.* Economic Strategy Institute. June 21, 1994. Washington, DC.

8. Amabile, Teresa. "Rethinking Rewards." *Harvard Business Review.* November–December, 1993.

9. Ibid.

10. McAdams, Jerry, "Rethinking Rewards." *Harvard Business Review.* November-December, 1993.

11. Shanley, Charles, "Testimony Presented to the United States House of Representatives, Committee on Small Business, Subcommittee on Regulation, Business Opportunities, and Energy." December 4, 1992, Washington, DC.

12. Rose, Kenneth H. "A Performance Measurement Model." Pacific Northwest Laboratory. Date unknown.

13. Rose.

14. *AFMC Technology Transfer Handbook.* Item F6.

15. Wolters, Donita S., "Rethinking Rewards." *Harvard Business Review.* November–December, 1993.

16. Hodge, Ron, Manager of Project Development, General Electric Corporate Development Center. Personal Interview. July 25, 1994.

Bibliography

"A Vision of Change for America," Executive Office of the President of the United States, U.S. Government Printing Office (ISBN 0-16-041662-0), Washington, D.C., February 17, 1993.

AFMC Technology Transfer Handbook, ASC/SMT, December 1993, distributed in 1994.

Amabile, Teresa. "Rethinking Rewards." *Harvard Business Review*. November–December, 1993.

Badaracco, Joseph, *The Knowledge Link: How Firms Compete Through Strategic Alliances*, McGraw, 1990.

Bagur, Jaques D. and Gusissinger, Ann S., "Technology Transfer Legislation: An Overview," *Journal of Technology Transfer*, Vol. 12, 1987, pp. 51–63.

Baron, Jonathan, "The Pilot Technology Access Program: A Federal Experiment in Technology Transfer," *Journal of Technology Transfer*, Vol. 15, 1990, pp. 25–30.

Battelle, "Not Invented Here: How U.S. Industry Sources Technology from the Public Sector," May 1994.

Benchmarking Techniques Workshop: A Two-Day Course, AT&T Benchmarking Group, AT&T, January 1992.

Berger, Beverly, "Technology Transfer in a Time of Transition: A Guide to Defense Conversion," Unpublished paper presented by the Federal Laboratory Consortium for Technology Transfer, January, 1994.

Berkowitz, Bruce D., "Can Defense Research Revive U.S. Industry?" *Issues in Science and Technology*, Winter, 1992–3, pp. 72–81.

Blanchard, P.A. and McDonald, C., "Reviving the Spirit of the Evolutionary Change in R&D Organizations," *Physics Today*, Vol. 39, pp. 42–50.

Bloch, E., *Toward a U.S. Technology Strategy*. Washington, D.C.: National Academy Press, 1991.

Bopp, G. (ed.), *Federal Lab Technology Transfer*. New York: Praeger Publishers, 1988.

Bozeman, Barry, *All Organizations are Public: Bridging Public and Private Organization Theory*, San Francisco, CA: Jossey-Bass, 1987.

Bozeman, Barry, "Evaluating Government Technology Transfer: Can the New 'Cooperative Technology Development' Policies Enhance U.S. Competitiveness?" Available from the author.

Bozeman, Barry and Coker, Karen, "Assessing the Effectiveness of Technology Transfer from U.S. Government R&D Laboratories: the Impact of Market Orientation," *Technovation*, Vol.12, No. 4, 1992, pp. 239–255.

Bozeman, Barry and Coursey, David, "A Typology of Industry-Government Laboratory Cooperative Research: Implications for Government Laboratory Policies and Competitiveness," Available from the authors.

Bozeman, Barry and Crow, Michael, "Technology Transfer from U.S. Government and University R&D Laboratories," *Technovation*, Vol. 11, No. 4, 1991. Pp. 231–246.

Bozeman, Barry and Crow, Michael, "The Environments of U.S. R&D Laboratories: Political and Market Influences," *Policy Sciences*, Vol. 23, 1990, pp. 25–56.

Bozeman, Barry and Fellows, Maureen, "Technology Transfer at the U.S. National Laboratory: A Framework for Evaluation," *Evaluation and Program Planning*, Vol. 11., 1988. Pp. 65–75.

Bozeman, Barry and Pandey, Sanjay, "Government Laboratories as a 'Competitive Weapon'—Comparing Cooperative R&D in the U.S. and Japan,," Paper through the Center for Technology and Information Policy, Smith College.

Bozeman, B. and Pandey, S., "Productivity Barriers in Public and Private R&D Laboratories: Effects of Sector, 'Publicness,Õ and Market Influence," paper through Center for Technology and Information Policy, 1993.

Bozeman, Barry, Rahm, Diane, and Crow, Michael, "Domestic Technology Transfer and Competitiveness: An Empirical Assessment of Roles of University and Governmental R&D Laboratories," *Public Administration Review*, November/December, 1988.

Branscomb, Lewis M., "National Laboratories: The Search for New Missions and New Structures," in Lewis M. Branscomb, ed., *Empowering Technology: Implementing a U.S. Strategy*. Cambridge, Mass.: MIT Press, 1993.

Calantone, R.J., Lee, M.T., and Gross, A.C., "Evaluating International Technology Transfer in a Comparative Marketing Framework," *Journal of Global Marketing*, Vol. 3, No. 5, 1990, pp. 20–40.

Camp, Robert C., *Benchmarking: The Search for Industry Best Practices that Lead to Superior Performance*, ASQC Quality Press, Milwaukee, WI, 1989.

Carr, R., "Doing Technology Transfer in Federal Laboratories," *Journal of Technology Transfer*, Spring–Summer 1992, pp. 8–23.

Chapman, R.L., "Implementing the 1986 Act: Signs of Progress," *Journal of Technology Transfer*, Vol. 14, 1989, pp. 5–13.

Chapman, R.L., Lohman, L.C., and Chapman, M.J., *An Exploration of the Benefits from NASA "Spinoff"*, Littleton, Co: Chapman Research Group.

Chiang, J., "From 'Mission–Oriented' to 'Diffusion–Oriented' Paradigm: The New Trend of U.S. Industrial Technology Policy," *Technovation*, Vol. 11, 1991, pp. 339–354.

Clinton, William J. and Gore, Albert, Jr., "Technology for America's Economic Growth: A New Direction to Build Economic Strength," from a Defense Symposium on Government, Industry, and Academia (Research): Partnership for a Competitive America, Fort Lesley J. McNair, Washington, D.C., April 7, 1993.

Council on Competitiveness, "Industry as a Customer of the Federal Laboratories," Washington, D.C.: Council on Competitiveness, 1993.

"Defense Acquisition Reform," The Report of the Defense Science Board Task Force on Acquisition Reform, Office of the Under Secretary of Defense for Acquisition and Technology, Washington, D.C., August 1994.

"Diffusing Innovations: Implementing the Technology Transfer Act of 1986". United States General Accounting Office, GAO/PEMD-91-23, May 1991.

"DoD Report Says New Measures Needed to Boost Technology Transfer," McGraw-Hill's Federal Technology Report, April 14, 1994.

Economic Strategy Institute, "Play to Win: Toward a New Paradigm for USAF Research and Development," A Defense Industry Strategy Proof-of-Concept Study, January 1993.

Freeman, C. *The Economics of Industrial Innovation*, MIT Press, Cambridge, MA, 1982.

Fusfeld, H., *The Technical Enterprise*, Cambridge, MA: Ballinger Press, 1986.

Gillespie, G.C., "Federal Laboratories: Economic Development and Intellectual Property Constraints," *Journal of Technology Transfer*, Vol. 13, 1988, pp. 20–26.

Government Accounting Office, *Case Study Evaluations*, Transfer Paper 9, April 1987.

Government Accounting Office, *Implementing the Technology Transfer Act of 1986*, May 1991.

Guile, B.R. and Quinn, J.B. (editors) *Managing Innovation: Cases from the Services Industries*, National Academy Press, Washington, D.C., 1988.

Hittle, Audie, "Technology Transfer Through Cooperative Research and Development," Massachusetts Institute of Technology, Sloan School's Management of Technology Program thesis. June 1991.

Hodge, Ron. Manager of Project Development, General Electric Corporate Development Center. Personal Interview. July 25, 1994.

"Improving the Air Force Technology Transfer Program: Best Current Processes and Alternative Approaches to Outstanding Issues" (Interim Technical Report), 1 August 1994. Prepared for the Directorate of Science and Technology, US Air Force Materiel Command, Wright-Patterson AFB, OH, 45433. Supportability Investment Decision Analysis Center (SIDAC), contract F33657-92-D-2055/0036, Battelle/SIDAC, 5100 Springfield Pike, Dayton, OH 45431.

Johngrud, C.C., Thornton, C.S., and Horak, T., "Technology Transfer: Attitude and Practices in U.S. Firms," *Management of Technology III: The Key to Global Competitiveness*, Proceedings from the Third International Conference on the Management of Technology, T.M. Khalil and B.A. Bayraktar, eds., Industrial Engineering and Management Press, Institute of Industrial Engineers; Norcross, GA, 1992, pp. 319–328.

Kearns, D., "Federal Labs Teem with R&D Opportunities," *Chemical Engineering*, Vol. 97, 1990, pp. 131–137.

Krieger, J.H., "Cooperation Key to U.S. Technology Remaining Competitive," *Chemical and Engineering News*, 1987.

Kuehn, T.J. and Porter, A.L. (editors) *Science, Technology, and National Policy*, Cornell University Press, Ithaca, NY, 1981.

Lee, J.W., "Improvement of Technology Transfer from Government Laboratories to Industry," Fifteenth Annual Meeting Proceedings, R.W. Harrison, ed., Technology Transfer Society, Dayton, OH, 1990.

Lesko, J.N. and Garvin, J.T. "The R&D Coordinator: Functioning as a Technical Liaison Officer," *Army Research, Development & Acquisition*, November–December 1989.

Lesko, J.N., Illinger, J.I, and Gonano, J.R. *The United States Army Advanced Concepts and Technology Program: Success Stories and a Case Study in Technology Transfer*, Government Printing Office, Washington, D.C., March 1993.

Lesko, J.N. and Reed, R. "Assessment of Technology Transfer Opportunities at the Technology 2003 Conference and Exposition," Battelle, Tactical Warfare Simulation and Technology Information Analysis Center (TWSTIAC) Report, contract DAAH01-88-C-0131, Columbus, OH, January 1994.

Lesko, J.N. and Schecter, G. "Retrospective Evaluation of the Army Advanced Concepts and Technology (ACT) Program," Battelle Final Report for the Army Research Office, Scientific Services Program, contract DAAL03-86-D-0001, DO 2646, Battelle, Arlington, VA, March 1993.

Link, A. and Bauer, L., *Cooperative Research in U.S. Manufacturing: Assessing Policy Initiatives and Corporate Strategies*, 1989.

Link, A. and Tassey, G. *Strategies for Technology-Based Competition*, Lexington, MA, Lexington Books, 1987.

McAdams, Jerry. "Rethinking Rewards." *Harvard Business Review*. November–December, 1993

McGregor, Douglas. "The Human Side of Enterprise." The Manager's Bookshelf: A Mosaic of Contemporary Views, Third Edition. Edited by Jon Pierce and John Newstrom. Harper Collins Publishers, 1993.

"Militarily Critical Technologies List," DoD, October 1992

Millett, S.M. and Honton, E.J. *A Manager's Guide to Technology Forecasting and Strategy Analysis Methods*, Battelle Press, Columbus, OH, 1991.

Morton, M.S. (editor) *The Corporation of the 1990s: Information Technology and Organizations Transformation* (Oxford University Press, New York, NY) 1991.

Mowery, David C. and Rosenburg, Nathan, *Technology and the Pursuit of Economic Growth*. Cambridge, England: Cambridge University Press, 1989.

"National Critical Technologies List," *Report of the National Critical Technologies Panel*, 1991

National Science Board, *Science & Engineering Indicators—1993*, Washington D.C., U.S. Government Printing Office, 1993.

Pages, Erik, "Next Steps in Business Conversion," Business Executives for National Security Special Report, April 1993.

"Partnering in Technology with the Federal Government: A Quick Reference", Oak Ridge Institute for Science and Education (ORISE). ORISE is managed by Oak Ridge Associated Universities for the US Department of Energy under contract DE-AC05-76OR00033, July 1993.

Pelz, DC and Andrews, F.M. *Scientists in Organizations: Productive Climates for R&D*, John Wiley & Sons, New York, NY, 1966.

Pinkston, J.T., "Technology transfer: issues for consortia," in K.D. Walters ed., *Entrepreneurial Management: New Technology and New Market Development*, Cambridge, MA, Ballinger Press, 1989, pp. 143–149.

Proceedings of Roundtable One. Economic Strategy Institute. June 21, 1994. Washington, DC.

"Quality: Small and Midsize Companies Seize the Challenge—Not a Moment too Soon,," *Business Weekly*, November 11, 1992.

"Re-engineering the Acquisition Oversight and Review Process," Volumes One & Two. Final Report to the Secretary of Defense by the Acquisition Reform Process Action Team, 9 December 1994.

Roberts, E.B. *The Dynamics of Research & Development*, Harper Row, New York, NY, 1964.

Roessner, J. David, "Technology Policy in the United States: Structures and Limitations," *Technovation*, Vol. 5, pp. 229–245.

Roessner, J. David, "What Companies Want from the Federal Labs," *Issues in Science and Technology,* Vol. X, No. 1, Fall 1993, p.37 ff.

Roessner, David J. and Bean,Alden, "Federal Technology Transfer: Industry Interactions with Federal Laboratories," *Journal of Technology Transfer*, Fall, 1990, pp. 5–14.

Roessner, David J. and Wise,Anne, "Patterns of Industry Interaction with Federal Laboratories," Atlanta, GA.: School of Public Policy, Georgia Institute of Technology, May, 1993.

Rose, Kenneth H. "A Performance Measurement Model." Pacific Northwest Laboratory. Date unknown.

Roussel, P.A. *et. al. Third Generation R&D: Managing the Link to Corporate Strategy*, Harvard Business School Press, Boston, MA, 1991.

Schacht, Wendy H., "Department of Energy Laboratories: A New Partnership with Industry?," from CRS Report, September 22, 1993.

Schacht, Wendy H., "Technology Transfer: Use of Federally Funded Research and Development," from CRS Report, March 30, 1994.

Schacht, Wendy H, "Cooperative R&D: Federal Efforts to Promote Industrial Competitiveness," from CRS Report, March 30, 1994.

Schriesheim, A., "Toward a Golden Age for Technology Transfer," *Issues in Science and Technology*, vol. II, 1990, pp. 52–58.

Shanley, Charles. "Testimony Presented to the United States House of Representatives, Committee on Small Business, Subcommittee on Regulation, Business Opportunities, and Energy." December 4, 1992, Washington, DC.

"Science in the National Interest," Executive Office of the President, Office of Science and Technology Policy, U.S. Government Printing Office (ISBN 0-16-045186-8), Washington, DC, August 1994.

Scott, William B., "NASA Reshapes Tech Transfer," *Aviation Week & Space Technology*, May 16, 1994, p.55

Shapira, Phillip, Roessner, David J., and Barke, Richard, "Federal-State Collaboration in Industrial Modernization," Atlanta, Ga.: School of Public Policy, Georgia Institute of Technology, 1992.

Shariq, Sayed, "Commercial Technology Mission: NASA Working for United States Economic Security," From Report of Office of Advanced Concepts and Technology, NASA, January 20, 1994.

Smilor, Raymond and Gibson, David, "Technology Transfer in Multi-Organizational Environments: The Case of R&D Consortia," *IEEE Transactions on Engineering Management*, Vol. 38, January, 1991, pp. 3–13.

Sobczak, T. "Technology Transfer: Who, What, Where & How," *Defense Electronics*, January 1995.

Soderstrom, E.J. and Winchell, B.M., "Patent Policy Changes Stimulating Commercial Application of Federal R&D," *Research Management*, Vol., 29, pp.35–38.

Soni, Som R., *Techtransfer & CRADA with Federal Laboratories*, LC 94-70437, AdTech Systems Research, Inc., 1994.

Souder, W.E., Nashar, A.S., and Padmanabhan, V. , "A Guide to the Best Technology Transfer Practices," *Journal of Technology Transfer*, Vol. 15, Nos. 1&2, 1990, pp. 5–16.

Spann, Mary S., Adams, Mel, and Sounder, William E., "Improving Federal Technology Commercialization: Some Recommendations From a Field Study," *Technology Transfer*, Summer–Fall, 1993, pp. 63–74.

Stewart, T.A. "Your Company's Most Valuable Asset: Intellectual Capital," *Fortune*, October 3, 1994.

"Survey of Laboratories and Implementation of the Federal Defense Laboratory Diversification Program," Director of Defense Research and Engineering. This report, dated February 1994, responds to Public Law 102-484, Section 4224(b), 23 October 1992. It includes annexes from each of the three military services: Annex A, Army (DTIC AD#-A277791); Annex B, Navy (AD-277792); and Annex C (AD-A277793), Air Force—each entitled "Domestic Technology Transfer."

Sutton, Jeanne, "Marrying Commercial and Military Technologies: Strategy for Maintaining Technological Supremacy," in *Essays on Strategy X,* Mary Sommerville, editor, NDU PRESS, April 1993

"Technology Transfer: Barriers Limit Royalty Sharing's Effectiveness," United States General Accounting Office, GAO/RCED-93-6, December 1992.

"Technology Transfer: Improving the Use of Cooperative R&D Agreements at DOE's Contractor-Operated Laboratories," United States General Accounting Office, GAO/RCED-94-91, April 1994.

Tyszkiewicz, M., "Research Collaboration in the European Community: Lessons for the U.S.," *Journal of Technology Transfer*, Fall, 1991, pp. 43–49.

United States General Accounting Office, *Diffusing Innovations: Implementing the Technology Transfer Act of 1986*, Washington, D.C.: USGPO, May, 1991.

Werner, J. and Bremer, J., "Hard Lessons in Cooperative Research," *Issues in Science and Technology*, Spring, 1991, pp. 44–49.

Wolters, Donita S. "Rethinking Rewards." *Harvard Business Review*. November–December, 1993.

Wyden, R., "Technology Transfer Obstacles in Federal Laboratories: Key Agencies Respond to Subcommittee Survey," Subcommittee on Regular Business Opportunities and Energy of the Committee on Small Business, House of Representatives, Washington, D.C., USGPO: Committee Print, 1990, pp. 101–103.

Index